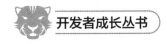

开发者成长丛书

# 仓颉语言实战

### 微课视频版

张 磊 ◎ 著

清华大学出版社

北京

## 内 容 简 介

本书是面向仓颉语言初学者的入门书，包括入门篇、进阶篇和高级篇，共计 27 章。

入门篇（第 1～7 章），目标是零基础入门仓颉语言。从最简单的 Hello World 示例开始，先讲解基本数据类型、变量、操作符等基础概念，然后讲解函数和流程控制，最后讲解一个综合应用示例，学习完本篇，读者就可以快速入门仓颉语言了。

进阶篇（第 8～22 章），目标是掌握仓颉语言的基础知识。本篇按照循序渐进的原则讲解了核心的仓颉语言概念，包括类、接口、枚举、泛型等，此外还讲解了常用的基础类库用法及异常处理、包管理等内容。学习完本篇，读者就掌握了基本的仓颉语言开发技能。

高级篇（第 23～27 章），目标是学习初步的企业级开发知识，包括函数的高级用法、文件处理及并发处理，最后还介绍了仓颉编译器和调试器。本篇对于企业级开发比较基础，但对于初学者，还是有一定难度的，掌握了本篇内容就可成为一个真正的仓颉语言开发者。

本书适合仓颉语言的初学者、高等院校计算机相关专业的学生，以及有一定开发经验，并希望快速学习仓颉软件开发的开发者、对自主可控编程语言感兴趣的爱好者阅读。

版权所有，侵权必究。举报：010-62782989，beiqinquan@tup.tsinghua.edu.cn。

图书在版编目(CIP)数据

仓颉语言实战：微课视频版/张磊著. —北京：清华大学出版社，2024.6
(开发者成长丛书)
ISBN 978-7-302-61659-7

Ⅰ. ①仓… Ⅱ. ①张… Ⅲ. ①程序语言－程序设计 Ⅳ. ①TP312

中国版本图书馆 CIP 数据核字(2022)第 145428 号

责任编辑：赵佳霓
封面设计：刘　键
责任校对：李建庄
责任印制：丛怀宇

出版发行：清华大学出版社
　　　　网　　址：https://www.tup.com.cn，https://www.wqxuetang.com
　　　　地　　址：北京清华大学学研大厦 A 座　　邮　编：100084
　　　　社　总　机：010-83470000　　　　　　　　邮　购：010-62786544
　　　　投稿与读者服务：010-62776969，c-service@tup.tsinghua.edu.cn
　　　　质量反馈：010-62772015，zhiliang@tup.tsinghua.edu.cn
　　　　课件下载：https://www.tup.com.cn,010-83470236
印 装 者：河北盛世彩捷印刷有限公司
经　　销：全国新华书店
开　　本：186mm×240mm　　印　张：21　　字　数：475 千字
版　　次：2024 年 7 月第 1 版　　　　　　　　印　次：2024 年 7 月第 1 次印刷
印　　数：1～2000
定　　价：89.00 元

产品编号：096504-01

# 前言
## PREFACE

二十多年前的1997年,作为一名计算机专业的大一新生,笔者第1次接触了编程语言。一开始学习的是汇编语言,后来逐步学习了其他语言,例如C、C++、Java等,还自学了Pascal等语言。毕业以后,工作中使用的语言主要追随行业趋势,先用了一段时间PowerScript,然后是Delphi,从2003年开始,重点用C♯,一直到2017年,最后全面转向Java。中间也接触了其他几种语言,例如Python、Scala等,但没有在工作中实际应用,也就没有深入研究。

但是,很遗憾,这些语言没有一种是我国自主研发的,在当前的大环境下,这是我们信息技术产业的又一个软肋,存在着被"卡脖子"的巨大风险,为了我国科技的发展,为了尽可能地减少不可预测事件的影响,我们有必要研发自己的编程语言。

众所周知,研发一种全新的语言不是一件容易的事情,语言本身的难度和创新性是一方面,配套的生态则是更巨大的挑战,所以说,大部分流行语言是业内有影响力的企业主持研发的,而社区主导的语言需要经过漫长的时间才有可能流行起来,可是,我们没有那么多时间。

幸运的是,华为编译器与编程语言实验室一直在做这件事情,并且有了重大的研发进展,但是,他们一直隐藏在幕后,直到2021年9月,华为邓泰华先生表示,将在2022年推出自研编程语言时,仓颉语言才进入了公众的视线。就个人来讲,最近几年,主要研究方向是鲲鹏领域,并在清华大学出版社出版了《鲲鹏架构入门与实战》,但是,对自主可控语言发展的关注一直没放松过,当仓颉语言项目的发起人兼项目经理王学智先生邀请我参加仓颉语言的内测时,我毫不犹豫地答应了,能参与仓颉语言的研发是我的荣幸,也是支持自主可控语言发展的方式之一。

在参加仓颉语言内测的同时,也根据测试的情况提交了一些仓颉语言的问题(Issue),在仓颉编程语言社区发布的2021年度报告中,被列为提交Issue最多的三个人之一。不过,坦率地说,仓颉语言毕竟是一种全新的语言,官方的资料只能逐步提供,所以,要推进仓颉生态的发展,就需要更多的参与者主动编写仓颉语言学习资料,为以后其他人的学习提供方便。在和清华大学出版社赵佳霓编辑沟通后,就决定编写一本面向仓颉语言初学者的入门书,也就是这本《仓颉语言实战(微课视频版)》。本书的定位是"零基础入门仓颉语言",分为入门篇、进阶篇和高级篇,特别针对初学者的特点,提供了大量的代码示例,并且给出了详细的注解，可以说达到了"手把手教初学者入门"的预期目的。

在当前的世界里，虽然科技领域的争端还在继续，但是，我们欣慰地看到，在 ICT 领域，国产自主可控技术得到了空前大发展，从硬件到软件，从底层的操作系统、数据库到各种应用中间件都有了很多颇具代表性的产品，这次华为推出的仓颉语言更是"向下扎到根，向上捅破天"的具体体现，我相信，以我们国家程序员的聪明才智，加上我们软件企业的规模和体量，仓颉语言一定能快速建立生态，成长为和 Java、Python 等著名语言齐名的新一代开发语言。

## 本书特色

本书践行"零基础入门仓颉语言"的核心理念，在章节设计和内容编排上具有以下几个特点。

（1）内容通俗易懂，使用平白的文字介绍仓颉开发的相关知识，在介绍语言特性时，不仅介绍特性本身，还会介绍为什么需要这种特性，能解决什么问题，帮助读者加深理解。

（2）知识点循序渐进，按照章节顺序学习即可，新知识点学习只依赖学过的内容，很少需要参考后续章节。

（3）丰富的代码示例，本书包含 200 多个完整的示例代码，每段代码都针对书中知识点精心设计，按照《CangJie 语言通用编程规范》编写，包括详细的代码注释和说明，并且每段代码均可以独立运行。

## 本书内容

第 1 章：仓颉语言简介，包括仓颉语言的发展经历及语言的特点。

第 2 章：仓颉开发准备，包括如何安装仓颉开发需要的工具链，以及编辑器、开发插件等。

第 3 章：第一个仓颉程序，从经典的 Hello World 开始，介绍了代码的编写、编译及程序的运行。

第 4 章：基本数据类型与操作符，包括标识符、关键字、变量等基础概念，以及仓颉语言基本数据类型和操作符，通过具体示例的方式，演示数据类型和操作符的使用方法。

第 5 章：函数，包括函数的定义、参数、返回值、函数体等内容，最后介绍了嵌套函数。

第 6 章：流程控制，包括条件表达式、循环表达式、match 表达式等。

第 7 章：入门综合实例，通过一个综合的入门实例，融会贯通前几章学习的知识，包括变量、函数、流程控制等内容，从总体上掌握基础的仓颉程序开发。

第 8 章：struct 类型，包括 struct 类型的必要性及如何定义和使用。

第 9 章：class 类型，包括对象、继承等关键概念，展示在仓颉语言里使用面向对象编程的方法。

第 10 章：enum 类型，重点介绍 enum 的构造器。该章是学习后续知识的基础。

第 11 章：接口，包括接口的必要性及接口的定义、实现、继承，最后介绍了仓颉语言内置的典型接口。

第 12 章：泛型，包括泛型的必要性及泛型接口、泛型函数和泛型约束的用法，最后介绍了泛型类型，包括泛型 class、泛型 struct、泛型 enum。

第 13 章：包管理，通过具体示例演示包的导入和导出。掌握了包的使用方法，就可以在开发中使用仓颉语言内置的大量库函数，也可以使用第三方提供的包。

第 14 章：扩展，演示如何在不破坏封装的情况下，给已有的类型添加成员或者实现接口。

第 15 章：基础集合类型，包括 Array 和 ArrayList，通过对比的方式展示在访问方式上的共性及在修改操作上的区别，从而可以更好地了解各自的适用场景。

第 16 章：函数的进阶用法，包括函数的重载，操作符的重载，以及函数作为"第一类对象"带来的特性。

第 17 章：类型关系，包括类型之间的关系及类型转换的方式，了解了类型关系就能更好地了解多态，了解面向对象的编程方式。

第 18 章：异常，包括异常的定义和异常处理的方法，重点介绍了 Option 类型在异常处理中的使用。

第 19 章：基础类库，包括格式化库、随机数库、数学库、时间库等，了解基础类库的常用用法，可以更好地进行程序编写。

第 20 章：字符及字符串处理，包括字符及字符串的十余种常用操作方法，通过一个游戏示例，演示字符串在开发中的实际应用。

第 21 章：高级集合类型，包括表示不重复集合的 HashSet 和表示键-值对的 HashMap，这两种类型在企业级开发中经常用到。

第 22 章：模式匹配，模式匹配的概念及 match 表达式支持的 6 种模式。

第 23 章：函数的高级用法，包括广泛使用的 Lambda 表达式及闭包和函数调用语法糖。

第 24 章：并发，包括仓颉语言特有的线程模型"仓颉线程"，同时介绍了解决并发编程数据同步问题的常用方式。

第 25 章：文件处理，包括文件、目录的基础操作和文件读写的实现。

第 26 章：仓颉编译器，包括常用编译选项及如何进行条件编译。

第 27 章：仓颉调试器，包括常用调试命令，通过实际示例的方式演示这些调试命令的具体用法。

## 致谢

感谢华为编译器与编程语言实验室的壮举，你们是新时代的开拓者！

感谢清华大学出版社,特别是以赵佳霓编辑为代表的工作人员,专业、严谨、细致的你们始终是我学习的榜样,也是本书顺利出版的保证。

感谢参与仓颉语言内测的所有朋友,和你们的交流使我对仓颉语言有了更深入的理解。

最后感谢我的妻子小朱同学和我们的孩子婉婉小朋友,在备战中考的同时还尽力给我提供了安心的写作环境,你们永远是我的骄傲。

作　者

2024 年 5 月于青岛

本书源代码

# 目 录
## CONTENTS

## 入 门 篇

### 第1章 仓颉语言简介 ············ 3
1.1 仓颉语言的由来 ············ 3
1.2 仓颉语言的特点 ············ 3

### 第2章 仓颉开发准备(▶ 12min) ············ 5
2.1 安装仓颉工具链 ············ 5
　2.1.1 Linux ············ 5
　2.1.2 Windows ············ 7
2.2 安装 VS Code 及仓颉插件 ············ 7
2.3 仓颉插件的使用 ············ 10
　2.3.1 仓颉项目结构 ············ 10
　2.3.2 语言插件的使用 ············ 10

### 第3章 第一个仓颉程序(▶ 6min) ············ 16
3.1 运行 Hello World 程序 ············ 16
3.2 仓颉程序基本规则 ············ 17
3.3 仓颉程序的编译 ············ 17

### 第4章 基本数据类型与操作符(▶ 21min) ············ 19
4.1 标识符与关键字 ············ 19
　4.1.1 标识符 ············ 19
　4.1.2 关键字 ············ 20
4.2 变量 ············ 20
4.3 基本数据类型 ············ 23
　4.3.1 整数类型 ············ 23

4.3.2　浮点类型 26
　　4.3.3　布尔类型 27
　　4.3.4　字符类型 28
　　4.3.5　字符串类型 29
　　4.3.6　Unit 类型 31
　　4.3.7　元组类型 31
　　4.3.8　区间类型 32
　　4.3.9　Noting 类型 33
4.4　基本数据类型转换 33
　　4.4.1　数值类型之间的转换 33
　　4.4.2　Char 和 UInt32 之间的转换 34
　　4.4.3　类型判断 34
4.5　操作符 35
　　4.5.1　算术操作符 35
　　4.5.2　逻辑操作符 38
　　4.5.3　位操作符 39
　　4.5.4　关系操作符 43
　　4.5.5　赋值操作符 44
　　4.5.6　操作符的优先级 45

# 第 5 章　函数（23min） 47

5.1　函数的定义 47
5.2　参数及函数调用 47
5.3　返回值类型 49
5.4　函数体 50
5.5　嵌套函数（局部函数） 51

# 第 6 章　流程控制（7min） 53

6.1　条件表达式 53
6.2　循环表达式 55
6.3　match 表达式 59

# 第 7 章　入门综合实例 62

7.1　开发需求 62
　　7.1.1　斐波那契数列 62
　　7.1.2　要解决的问题 62

7.2 解决思路 ································································································ 62
   7.2.1 问题分析 ·························································································· 62
   7.2.2 递归函数 ·························································································· 63
7.3 示例代码 ································································································ 63

## 进 阶 篇

### 第 8 章 struct 类型（19min） ···················································· 67

8.1 长方体引发的思考 ·················································································· 67
8.2 struct 类型的定义 ··················································································· 69
8.3 成员变量 ································································································ 70
8.4 构造函数 ································································································ 71
   8.4.1 普通构造函数 ···················································································· 71
   8.4.2 主构造函数 ······················································································· 72
   8.4.3 自动生成的无参构造函数 ································································· 72
8.5 成员函数 ································································································ 73
8.6 可见修饰符 ···························································································· 73
8.7 实例的创建与访问 ·················································································· 75
8.8 mut 函数 ································································································ 76
8.9 成员属性 ································································································ 76
   8.9.1 属性的定义 ······················································································· 77
   8.9.2 属性的使用 ······················································································· 78

### 第 9 章 class 类型（12min） ······················································ 80

9.1 定义 ······································································································· 80
9.2 成员变量 ································································································ 81
9.3 构造函数 ································································································ 81
   9.3.1 普通构造函数 ···················································································· 81
   9.3.2 主构造函数 ······················································································· 82
   9.3.3 自动生成的无参构造函数 ································································· 83
9.4 成员函数 ································································································ 83
9.5 成员属性 ································································································ 84
9.6 可见性修饰符 ························································································· 84
9.7 对象 ······································································································· 85
   9.7.1 对象的创建与访问 ············································································ 85
   9.7.2 对象值的修改 ··················································································· 86

9.8 抽象类 ⋯⋯ 89
9.9 继承 ⋯⋯ 89
 9.9.1 继承的定义 ⋯⋯ 89
 9.9.2 覆盖和重定义 ⋯⋯ 91
 9.9.3 super 关键字 ⋯⋯ 92
 9.9.4 成员可见性 ⋯⋯ 93

## 第 10 章 enum 类型 ⋯⋯ 96

10.1 enum 类型的定义 ⋯⋯ 96
10.2 enum 类型的值 ⋯⋯ 97
10.3 enum 类型的使用 ⋯⋯ 97
10.4 有参构造器 ⋯⋯ 98

## 第 11 章 接口（10min）⋯⋯ 99

11.1 为什么需要接口 ⋯⋯ 99
11.2 接口的定义 ⋯⋯ 99
11.3 接口的实现 ⋯⋯ 100
 11.3.1 接口的通常实现 ⋯⋯ 100
 11.3.2 接口的默认实现 ⋯⋯ 102
11.4 接口的继承 ⋯⋯ 103
11.5 类型的多接口实现 ⋯⋯ 104
11.6 典型的内置接口 ⋯⋯ 105
 11.6.1 Any 类型 ⋯⋯ 105
 11.6.2 ToString 接口 ⋯⋯ 105

## 第 12 章 泛型（17min）⋯⋯ 106

12.1 什么是泛型 ⋯⋯ 106
12.2 泛型接口 ⋯⋯ 108
12.3 泛型函数 ⋯⋯ 109
12.4 泛型约束 ⋯⋯ 110
12.5 泛型类型 ⋯⋯ 113
 12.5.1 泛型 class ⋯⋯ 113
 12.5.2 泛型 struct ⋯⋯ 114
 12.5.3 泛型 enum ⋯⋯ 115
 12.5.4 区间类型 ⋯⋯ 117

## 第13章　包管理 …… 118

### 13.1　包的声明 …… 118
### 13.2　顶层声明的可见性 …… 119
### 13.3　包的导出和编译 …… 119
### 13.4　包的导入 …… 120
#### 13.4.1　import 语句导入 …… 120
#### 13.4.2　导入重命名 …… 122

## 第14章　扩展（6min） …… 123

### 14.1　扩展的定义 …… 123
### 14.2　泛型扩展 …… 125
### 14.3　接口扩展 …… 126

## 第15章　基础集合类型（8min） …… 128

### 15.1　Array …… 128
#### 15.1.1　Array 的定义 …… 128
#### 15.1.2　访问 Array …… 128
#### 15.1.3　修改 Array …… 131
#### 15.1.4　Array 的高级用法 …… 132
#### 15.1.5　字节数组字面量 …… 134
### 15.2　ArrayList …… 134
#### 15.2.1　ArrayList 的定义 …… 134
#### 15.2.2　访问 ArrayList …… 134
#### 15.2.3　修改 ArrayList …… 135

## 第16章　函数的进阶用法（20min） …… 137

### 16.1　函数重载 …… 137
#### 16.1.1　函数重载的定义 …… 137
#### 16.1.2　函数重载决议 …… 139
### 16.2　函数遮盖 …… 141
### 16.3　操作符重载 …… 142
#### 16.3.1　操作符重载的必要性 …… 142
#### 16.3.2　操作符重载的定义 …… 145
#### 16.3.3　索引操作符 …… 146

16.3.4　可以被重载的操作符 ······ 147
16.4　函数是第一类对象 ······ 147
　　16.4.1　什么是第一类对象 ······ 147
　　16.4.2　函数类型的定义 ······ 148
　　16.4.3　函数作为参数 ······ 149
　　16.4.4　函数作为变量 ······ 149
　　16.4.5　函数作为返回值 ······ 150

# 第17章　类型关系 ······ 152

17.1　多态 ······ 152
17.2　子类型关系 ······ 154
　　17.2.1　继承带来的子类型关系 ······ 154
　　17.2.2　实现接口带来的子类型关系 ······ 154
　　17.2.3　元组类型的子类型关系 ······ 154
　　17.2.4　函数类型的子类型关系 ······ 154
　　17.2.5　预设子类型关系 ······ 156
　　17.2.6　传递性带来的子类型关系 ······ 156
　　17.2.7　泛型类型的子类型关系 ······ 156
17.3　类型转换 ······ 157
　　17.3.1　is 操作符 ······ 157
　　17.3.2　as 操作符 ······ 158
17.4　类型别名 ······ 159

# 第18章　异常 ······ 161

18.1　异常的定义 ······ 161
18.2　异常处理 ······ 162
18.3　自定义异常 ······ 163
18.4　Option 值的解构 ······ 164
　　18.4.1　getOrThrow()函数 ······ 164
　　18.4.2　?? 操作符 ······ 164

# 第19章　基础类库 ······ 166

19.1　格式化库 ······ 166
　　19.1.1　整型、浮点型类型 ······ 166
　　19.1.2　字符类型 ······ 171

## 19.2 Console 类 ······ 171
### 19.2.1 ConsoleReader ······ 172
### 19.2.2 ConsoleWriter ······ 175
## 19.3 Random 类 ······ 176
## 19.4 数学库 ······ 178
### 19.4.1 常数 ······ 179
### 19.4.2 函数 ······ 181
## 19.5 转换库 ······ 184
## 19.6 base64 包 ······ 186
## 19.7 hex 包 ······ 187
## 19.8 时间库 ······ 189
### 19.8.1 Month 枚举 ······ 189
### 19.8.2 DayOfWeek 枚举 ······ 191
### 19.8.3 Duration 类 ······ 192
### 19.8.4 TimeZone ······ 194
### 19.8.5 DateTime ······ 196
### 19.8.6 时间格式 ······ 201

# 第 20 章 字符及字符串处理 ······ 207

## 20.1 字符处理 ······ 207
## 20.2 字符串处理 ······ 208
### 20.2.1 字符串转数组 ······ 208
### 20.2.2 指定位置字符获取 ······ 209
### 20.2.3 子字符串获取 ······ 209
### 20.2.4 字符查找 ······ 209
### 20.2.5 子字符串查找 ······ 211
### 20.2.6 字符串修整 ······ 212
### 20.2.7 字符串分隔 ······ 213
### 20.2.8 字符串判断 ······ 215
### 20.2.9 字符串连接 ······ 215
### 20.2.10 字符串替换与反转 ······ 218
## 20.3 猜数字小游戏 ······ 218

# 第 21 章 高级集合类型（ 10min） ······ 223

## 21.1 Hashable 接口 ······ 223

21.2 HashSet 集合 ·········································································· 223
  21.2.1 HashSet 的定义 ············································································ 223
  21.2.2 访问 HashSet ················································································ 224
  21.2.3 修改 HashSet ················································································ 224
  21.2.4 HashSet 的容量和元素个数 ······························································ 227
21.3 HashMap 集合 ········································································· 229
  21.3.1 HashMap 的定义 ············································································ 229
  21.3.2 访问 HashMap ··············································································· 229
  21.3.3 修改 HashMap ··············································································· 230
  21.3.4 其他常用函数 ················································································ 231
  21.3.5 综合应用示例 ················································································ 232

## 第 22 章 模式匹配 ······································································ 235

22.1 match 表达式 ·········································································· 235
  22.1.1 pattern guard ················································································ 235
  22.1.2 match 表达式类型 ········································································· 236
22.2 模式 ······················································································ 237
  22.2.1 常量模式 ······················································································ 237
  22.2.2 通配符模式 ··················································································· 238
  22.2.3 变量模式 ······················································································ 238
  22.2.4 元组模式 ······················································································ 239
  22.2.5 类型模式 ······················································································ 239
  22.2.6 枚举模式 ······················································································ 240

# 高 级 篇

## 第 23 章 函数的高级用法（19min） ··················································· 245

23.1 Lambda 表达式 ········································································ 245
  23.1.1 Lambda 表达式的定义 ···································································· 245
  23.1.2 Lambda 表达式的返回值 ································································· 246
  23.1.3 Lambda 表达式的调用 ···································································· 247
23.2 闭包 ······················································································ 248
  23.2.1 闭包的定义 ··················································································· 248
  23.2.2 捕获变量的状态 ············································································ 249

23.2.3 可变变量的闭包 ………………………………………………… 250

23.3 函数调用语法糖 ………………………………………………………… 251

    23.3.1 尾随闭包 …………………………………………………………… 251

    23.3.2 管道表达式 ………………………………………………………… 253

    23.3.3 组合操作符 ………………………………………………………… 254

## 第 24 章 并发（ 18min） ………………………………………………… 256

24.1 仓颉线程 ………………………………………………………………… 256

    24.1.1 线程睡眠函数 sleep ……………………………………………… 256

    24.1.2 创建仓颉线程 ……………………………………………………… 257

    24.1.3 等待线程结束并获取返回值 ……………………………………… 258

24.2 同步 ……………………………………………………………………… 260

    24.2.1 数据竞争 …………………………………………………………… 260

    24.2.2 原子操作 …………………………………………………………… 262

    24.2.3 互斥锁 ……………………………………………………………… 267

    24.2.4 监视器 ……………………………………………………………… 269

    24.2.5 synchronized 关键字 ……………………………………………… 272

## 第 25 章 文件处理 ………………………………………………………… 275

25.1 FileInfo ………………………………………………………………… 275

25.2 File ……………………………………………………………………… 278

25.3 Directory ………………………………………………………………… 284

25.4 文件读写示例 …………………………………………………………… 288

## 第 26 章 仓颉编译器（ 14min） ………………………………………… 292

26.1 编译演示代码 …………………………………………………………… 292

26.2 编译选项 ………………………………………………………………… 293

26.3 条件编译 ………………………………………………………………… 299

    26.3.1 使用方式 …………………………………………………………… 299

    26.3.2 内置编译条件 ……………………………………………………… 300

    26.3.3 自定义编译条件 …………………………………………………… 301

    26.3.4 多条件编译 ………………………………………………………… 302

## 第 27 章 仓颉调试器（ 26min） ………………………………………… 303

27.1 仓颉调试器演示代码 …………………………………………………… 303

27.2 调试版本的编译 …………………………………………………………… 304
27.3 启动调试的方式 …………………………………………………………… 305
27.4 调试命令 …………………………………………………………………… 306
    27.4.1 断点 ………………………………………………………………… 306
    27.4.2 观察点 ……………………………………………………………… 309
    27.4.3 启动 ………………………………………………………………… 310
    27.4.4 执行 ………………………………………………………………… 310
    27.4.5 变量 ………………………………………………………………… 313
    27.4.6 退出 ………………………………………………………………… 317

# 入 门 篇

# 第 1 章 仓颉语言简介

## 1.1 仓颉语言的由来

编程语言的出现已经有很长一段时间了，即使从真正的高级编程语言 FORTRAN 正式发布算起（1956 年 10 月 15 日），距今也将近 70 年了。这些年来出现的编程语言有上千种，现在规模使用的编程语言也在百种以上，遗憾的是，这些语言中没有一种是由中国完全自主开发的。对于 2019 年以前的开发者来讲，并没有感觉到有什么不妥，毕竟大部分人喜欢学习最流行的编程语言，使用同一种语言还可以提高开发效率，降低沟通成本，单纯从技术角度来分析，这是最优的做法。

然而，复杂的现实世界出现了人为的技术壁垒，发展自主可控的编程语言成了我们必须克服的困难。华为公司在很早以前就开始了编译器的研究开发工作，在 2019 年 8 月发布并开源了方舟编译器；华为编程语言实验室也在同年发起了研发自主编程语言的项目，后期又邀请到了著名的编程语言专家冯新宇教授加盟；2021 年 2 月，华为注册了"仓颉语言"商标，2021 年 9 月，华为高管邓泰华表示将在 2022 年推出仓颉编程语言。然而，有点遗憾的是，2022 年并没有推出仓颉语言，而是把推出时间延长到了 2023 年秋季的全连接大会；后来，为了慎重起见，也为了打造更好的仓颉生态，仓颉语言再一次被推迟到了 2024 年上半年。

## 1.2 仓颉语言的特点

仓颉语言的研发也是信息产业发展的需要，在当前多设备、多应用、全场景的开发中，现有的编程语言并不能很好地适应这种变化，而仓颉语言作为一种全新的语言，具有显著的后发优势，可以兼收并蓄各种前代语言的优点，解决在实际应用中出现的问题，从而更好地为开发服务。仓颉语言的主要特点如下。

**1. 内置的高安全性**

仓颉语言是一种静态强类型语言，在编译期间可以发现程序中的类型错误，从而尽可能地减少运行时的异常；对于强类型语言来讲，一般需要在开发时就标注好类型，但是仓颉编译器具有强大的推断能力，可以根据上下文自动推断出类型，不单单是变量，即使是函数的返回值类型也可以推断出来，从而大大减少了开发者类型标注的工作量。

仓颉语言的垃圾收集机制可以自动进行内存管理，在运行时检查数组是否下标越界、是否溢出，保证了程序的安全可靠运行。

### 2. 支持多范式编程

一般来讲，常用的编程范式有 3 种，包括最早的面向过程的命令式编程，后期发展壮大成为最受欢迎的面向对象编程，以及现在如日中天的函数式编程。这几种编程范式各有优缺点，现在流行的语言对前两种大都有较好的支持；函数式编程虽然出现得很早，但是基本只在 Lisp 等几种研究性语言中使用，在面向企业级开发的语言中应用较晚，这就导致部分语言对函数式编程的支持是通过打补丁实现的。仓颉语言在设计之初就考虑了这几种编程范式，从根本上实现了对这几种范式的内置支持，从而可以更简洁、更高效地实现多范式编程。

### 3. 易用的并发、分布式编程

现代的处理器性能越来越强大，核数也越来越多，几十个核心甚至上百个核心的处理器也都出现并得到了大规模应用，为了充分利用这些核心的计算资源，仓颉语言提供了原生的用户态轻量化线程，易于开发、实现高并发的编程。

### 4. 跨语言的兼容性

作为一种新生的语言，最困难的地方在于生态及对第三方库的支持，仓颉语言也充分考虑了这一点，实现了对多语言的兼容，通过高效调用其他主流编程语言，进而实现对这些语言库的复用，从而解决语言初期的生态困境。目前已经实现了对 C 语言的跨语言支持，根据规划，还会实现对 Java 语言的跨语言支持。

### 5. 易扩展为领域专用语言

随着移动应用、人工智能、区块链、量子计算等领域的蓬勃发展，软件在各行各业的应用得到爆发式的增长，各行业的研究人员和领域专家也加入了软件开发的队伍中，这些领域专家对擅长的领域有非常深入的研究，但是没有计算机编程的基础，他们希望有一种屏蔽了计算机底层复杂度的简单语言，可以针对自己的领域进行编程，这种语言就是领域特定语言(DSL)。

如果开发一种全新的独立 DSL，则成本较高，并且需要打造完整的工具链，更优的做法是在现有的通用语言(例如 Java、Swift、仓颉)基础上开发，可以兼容现有的工具链，这种基于通用语言开发的领域专用语言被称为内嵌式领域专用语言(eDSL)。仓颉语言的高阶函数、尾随闭包、属性机制、操作符重载、部分关键字可省略等特性，有利于 eDSL 的构建。此外，仓颉语言还提供了基于宏的元编程支持，在编译时生成或改变代码，让开发者可以深度定制程序的语法和语义，构建更加符合领域抽象的语言特性。

### 6. 对人工智能(AI)开发的内置支持

人工智能是今后软件开发的重要发展方向，仓颉语言主要通过两种方式提供了对人工智能开发的支持，一种是语言级别的支持，主要体现在仓颉语言的自动微分等特性上；另一种是系统库级别的支持，包括内置的大量算子及仓颉高性能机器学习库。这些内在的支持，降低了人工智能开发的难度，从而为高效、快速的 AI 应用开发提供了基础。

# 第 2 章 仓颉开发准备

## 2.1 安装仓颉工具链

7min

在仓颉工具链中,最重要的工具之一是仓颉编译器,在本书编写时,仓颉编译器已适配 Linux 和 Windows 平台,下面分别进行介绍在这两个平台上的仓颉工具链安装。

### 2.1.1 Linux

仓颉工具链支持多种 Linux 发行版,不同架构下的环境要求如表 2-1 所示。

表 2-1 环境要求

| 架 构 | 环 境 要 求 |
| --- | --- |
| x86_64 | glibc 2.22 |
|  | Linux Kernel 4.12 或更高版本 |
|  | libstdc++ 6.0.24 或更高版本 |
| aarch64 | glibc 2.27 |
|  | Linux Kernel 4.15 或更高版本 |
|  | libstdc++ 6.0.24 或更高版本 |

本节选择的示例为 Linux Ubuntu 18.04 版本,下面演示在该环境下安装仓颉工具链的步骤。

步骤 1:登录 Ubuntu 服务器,查看版本信息。命令及可能的回显如下:

```
#lsb_release -a
LSB Version:
    core-9.20170808Ubuntu1-noarch:security-9.20170808Ubuntu1-noarch
Distributor ID:    Ubuntu
Description:    Ubuntu 18.04.6 LTS
Release:    18.04
Codename:    bionic
```

从回显中可以确认服务器操作系统版本为 Ubuntu 18.04。

步骤2(可选)：仓颉工具链依赖如下的软件包。
- Binutils
- libc-dev
- libc++-dev
- libgcc-7-dev

如果这些软件包不存在,则可以通过如下命令安装：

```
apt-get install binutils libc-dev libc++-dev libgcc-7-dev
```

步骤3(可选)：仓颉工具链依赖OpenSSL3。如果环境中不存在该依赖,则可以通过以下方式编译安装：

(1) 从 OpenSSL 官方网站 https://www.openssl.org/source/或者其他地址下载 OpenSSL3 的源码,建议使用 openssl-3.0.7 或更高版本,本示例使用 3.0.10 版本,下载网址为 https://www.openssl.org/source/openssl-3.0.10.tar.gz。

(2) 下载完毕,解压压缩包,命令如下：

```
tar -zxvf openssl-3.0.10.tar.gz
```

(3) 进入 openssl-3.0.10 目录,编译 OpenSSL,编译过程可能需要几分钟到十几分钟,命令如下：

```
./Configure --libdir=lib
make
```

(4) 测试 OpenSSL,检查编译结果是否正常,命令如下：

```
make test
```

(5) 安装 OpenSSL 到系统目录,命令如下：

```
make install
```

(6) 检查安装版本信息,命令及回显如下：

```
openssl version
OpenSSL 3.0.10 1 Aug 2023 (Library: OpenSSL 3.0.10 1 Aug 2023)
```

至此 OpenSSL3 安装完毕。

步骤4：创建/soft目录并进入。命令如下：

```
mkdir /soft
cd /soft/
```

步骤5：下载仓颉工具链安装包并解压。本演示使用的是 0.39.7 版本,这里假设已经下载了 x86 架构下的 Cangjie-0.39.7-linux_x64.tar.gz 软件包,解压命令如下：

```
tar -zxvf Cangjie-0.39.7-linux_x64.tar.gz
```

如果安装环境是 aarch64 架构，则可以安装对应架构的软件包，例如 Cangjie-0.39.7-linux_aarch64.tar.gz。

步骤 6：配置环境变量。编辑/etc/profile 配置文件，命令如下：

```
vim /etc/profile
```

在文件的最后一行下面加上如下的内容：

```
export CANGJIE_HOME = /soft/cangjie

export PATH = ${CANGJIE_HOME}/bin:${CANGJIE_HOME}/tools/bin:$PATH
export LD_LIBRARY_PATH = ${CANGJIE_HOME}/runtime/lib/linux_x86_64_llvm:${CANGJIE_HOME}/third_party/llvm/lldb/lib:${LD_LIBRARY_PATH}
```

然后保存退出。如果仓颉工具链不是安装在/soft/cangjie 目录下，则需要根据实际安装位置修改配置文件内容。

步骤 7：使环境变量生效。命令如下：

```
source /etc/profile
```

步骤 8：检查编译器配置是否正确。命令及成功的回显如下：

```
cjc - v
Cangjie Compiler: 0.39.7 (4aeaaf53492f 2023 - 08 - 24)
```

这样就完成了仓颉工具链的安装配置。

### 2.1.2 Windows

在 Windows 平台下，仓颉语言提供了两种安装包，分别是 exe 和 zip 格式，对于 exe 格式的安装包，按照向导提示安装即可；对于 zip 格式的安装包，在解压到适当目录后，执行 envsetup.bat、envsetup.ps1 和 envsetup.sh 三种安装脚本中的一个即可。

## 2.2 安装 VS Code 及仓颉插件

本书编写时，仓颉专用 IDE 即 CangjieStudio 尚未正式发布，读者可以使用 VS Code 作为编辑器进行仓颉程序的开发。接下来将演示在 Windows 平台上安装 VS Code 和仓颉插件的过程，其他平台安装方式类似。仓颉插件要求 VS Code 版本为 1.67 及以上，本次演示使用的是 1.82 版本。

步骤 1：进入 VS Code 1.82 版本下载网页 https://code.visualstudio.com/updates/v1_82，选择 System 版本下载并安装包，如图 2-1 所示。

步骤 2：按照安装包提示安装 VS Code。

步骤 3：解压插件包，以 0.39.7 为例，插件包为 Cangjie-vscode-0.39.7.tar.gz，解压该包可以得到插件文件 Cangjie-0.39.7.vsix。

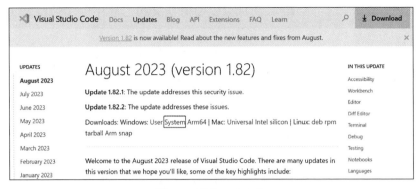

图 2-1　下载 VS Code

步骤 4：启动 VS Code，单击"扩展"，然后单击扩展栏的"…"图标，从弹出的菜单里单击"从 VSIX 安装"按钮，如图 2-2 所示。

图 2-2　本地安装插件

步骤 5：从插件所在的目录选择插件 Cangjie-0.39.7.vsix 进行安装，如图 2-3 所示。

图 2-3　选择安装的插件

安装好的插件可以从已安装扩展中看到,如图 2-4 所示。

图 2-4　已安装的插件

步骤 6:单击仓颉插件右下角的配置图标,弹出配置菜单,如图 2-5 所示。

图 2-5　配置插件

步骤 7:选择"扩展配置"选项,弹出仓颉 Sdk 路径配置窗口,如图 2-6 所示。

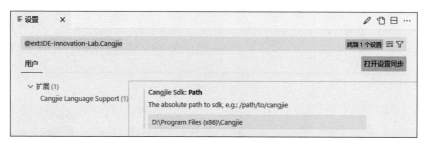

图 2-6　Sdk 路径配置

在输入框里输入仓颉 Sdk 的路径,本演示示例的路径为 D:\Program Files（x86）\Cangjie,这样,仓颉插件的配置就完成了。

## 2.3 仓颉插件的使用

### 2.3.1 仓颉项目结构

根据仓颉的要求,仓颉项目代码文件默认存放在 src 目录下,一个标准的项目结构如图 2-7 所示。

如果要把源码文件放到其他目录,则可以在工程目录下新建 modules.json 文件,在该文件里配置源码目录路径。假如源码存放在工程目录下的 other_src 目录,则 modules.json 可以如图 2-8 所示进行配置。

图 2-7 标准项目结构

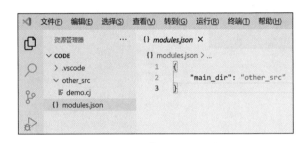

图 2-8 自定义源码目录

### 2.3.2 语言插件的使用

仓颉语言插件安装成功后,既可以在 VS Code 中使用它提供的多种增强功能。

**1. 语法高亮**

对于仓颉代码文件,插件自动执行语法高亮,如图 2-9 所示。

```
//根据操作符号和数字执行操作
func dealOpAndNum(opNum: OpAndNum) {
    askTimes++
    match (opNum.op) {
        case "==" | "=" => dealWithEqual(opNum.num)
        case ">" => PrintJudgmentResults(guessedNumber > opNum.num)
        case ">=" => PrintJudgmentResults(guessedNumber >= opNum.num)
        case "<" => PrintJudgmentResults(guessedNumber < opNum.num)
        case "<=" => PrintJudgmentResults(guessedNumber <= opNum.num)
        case _ => println("No")
    }
}
```

图 2-9 语法高亮

**2. 自动补全**

对于关键字、变量或者"."符号,在光标右侧显示候选内容,可以通过上下键快速选择,如图 2-10 所示。

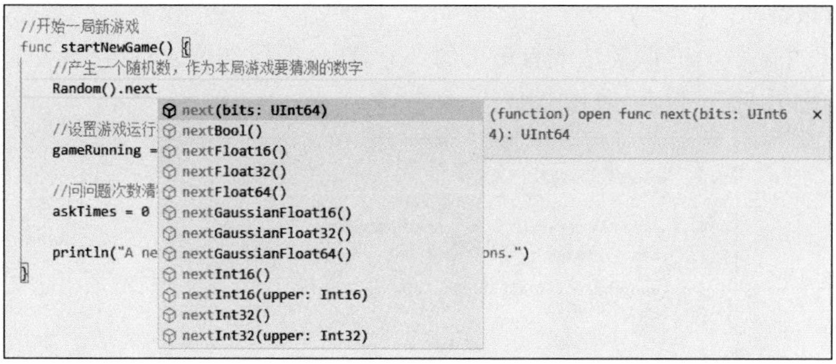

图 2-10　自动补全

### 3．定义跳转

将鼠标悬停在目标上然后按住 Ctrl 键并单击鼠标左键,可以转到目标定义位置,也可以在目标上右击,在弹出的菜单中选择"转到定义",或者通过快捷键 F12 也可以转到目标定义位置,如图 2-11 所示。

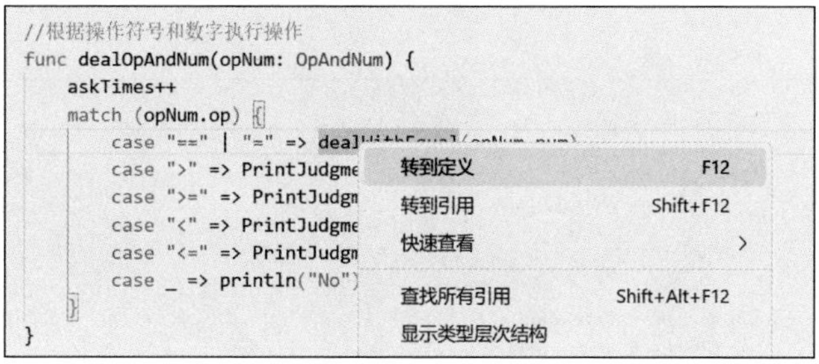

图 2-11　定义跳转

### 4．查找引用

在目标符号位置右击,在弹出的菜单里选择"查找所有引用",插件会查找所有对该符号的引用,如图 2-12 所示。

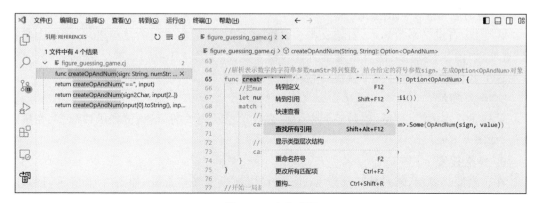

图 2-12　查找引用

在查找结果条目上单击,既可以自动跳转到对应的代码位置。

**5. 诊断报错**

对于代码中不符合仓颉语法或者语义的部分,插件会用红色波浪线标示出来,鼠标悬停在上面会出现错误提示信息,如图2-13所示。

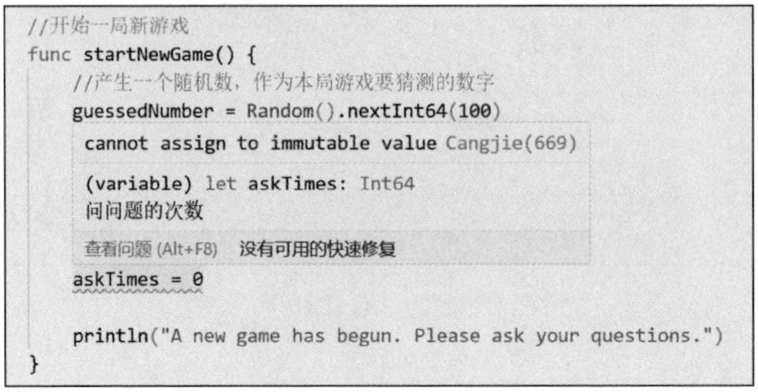

图 2-13　诊断报错

**6. 选中高亮**

当选中一个变量或者函数时,在定义处和所有使用该变量或者函数的地方都会高亮显示,如图2-14所示。

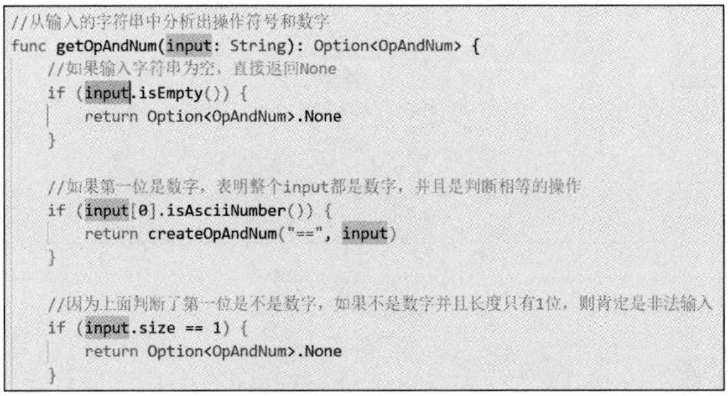

图 2-14　选中高亮

**7. 重命名**

在变量、自定义函数名称、自定义类型名称上右击,然后在弹出的菜单里选择"重命名符号"选项,如图2-15所示,即可出现重命名的输入界面,如图2-16所示,输入新的名称,所有使用这个名称的地方都会被替换成新名称。

**8. 悬浮提示**

鼠标悬停在类型、变量、函数等元素上面会自动提示相关的信息,例如函数原型、变量定义等,如图2-17所示。

**9. 定义搜索**

在仓颉源码界面,通过按快捷键Ctrl+T可以弹出定义搜索框,如图2-18所示,输入匹

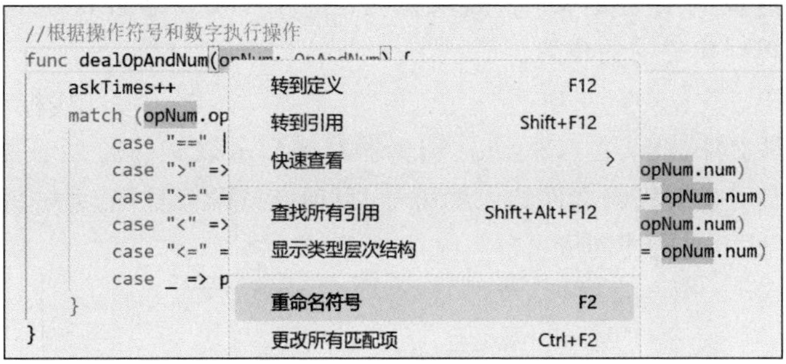

图 2-15　重命名菜单

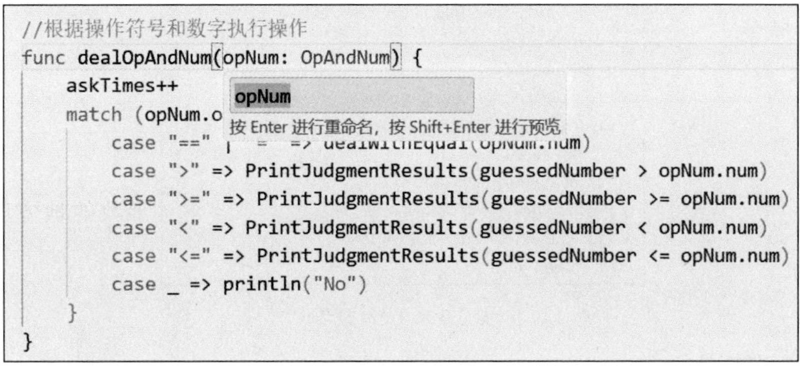

图 2-16　重命名

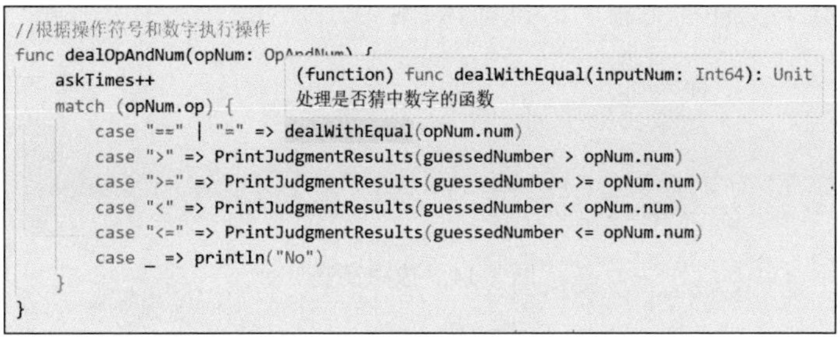

图 2-17　悬浮提示

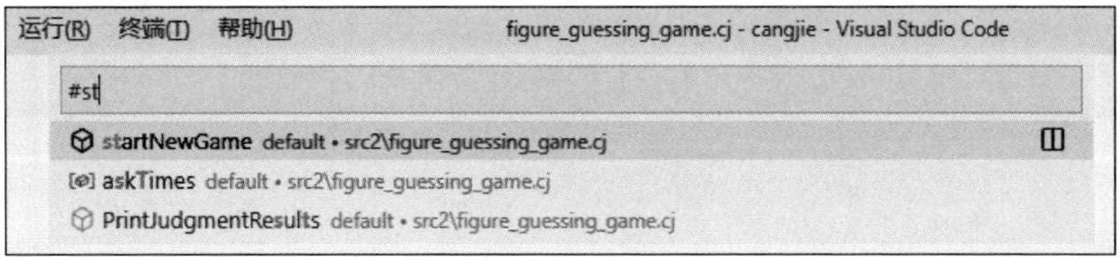

图 2-18　定义搜索

配的名称可以快速找到类型定义,在结果条目上单击,或者通过键盘选中条目,然后按Enter键,将可以转到代码中的位置。

**10. 大纲视图**

对于仓颉代码文件,默认在 VS Code 编辑器左侧显示该文件的大纲视图,如图 2-19 所示。大纲视图列出了仓颉代码文件的主要符号,目前支持两级呈现,第一级显示顶层定义的声明,第二级主要显示构造器和成员。

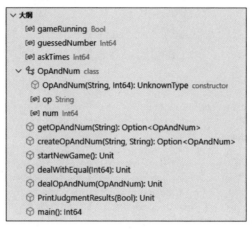

图 2-19 大纲视图

**11. 面包屑导航**

对于项目中的仓颉代码文件,插件支持面包屑导航,在编辑界面的左上角,显示了该符号所处的位置和代码文件在工程中的位置,如图 2-20 所示。

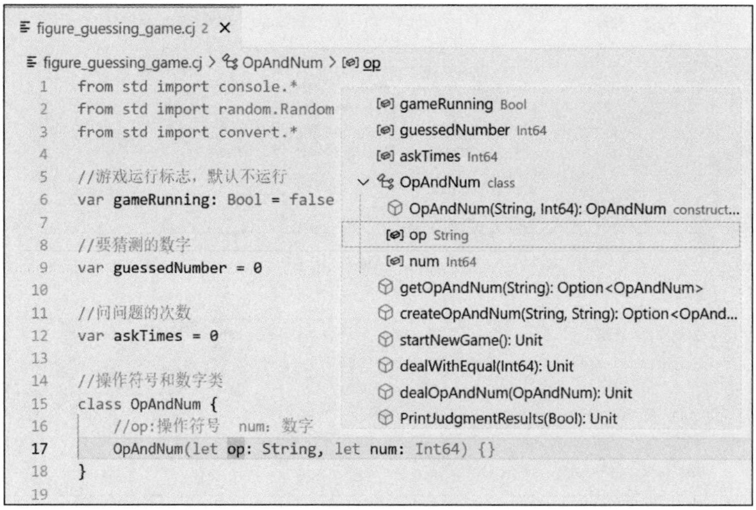

图 2-20 面包屑导航

## 12．签名帮助

在代码编辑界面,左括号和逗号会触发函数的签名帮助,给开发者提供当前函数的参数信息,如图 2-21 所示。

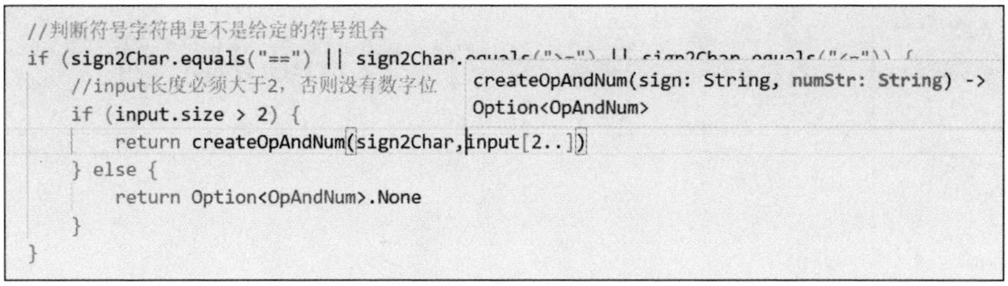

图 2-21　签名帮助

## 13．格式化

右击仓颉代码文件编辑界面,在弹出的菜单中选择 Format 选项,如图 2-22 所示,将会触发代码格式化功能,格式化后的代码如图 2-23 所示,代码的可读性有了显著提高。

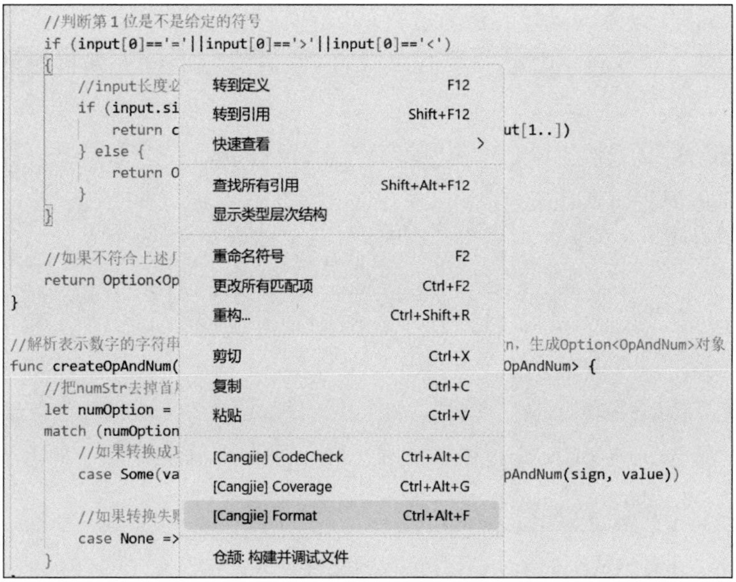

图 2-22　格式化菜单

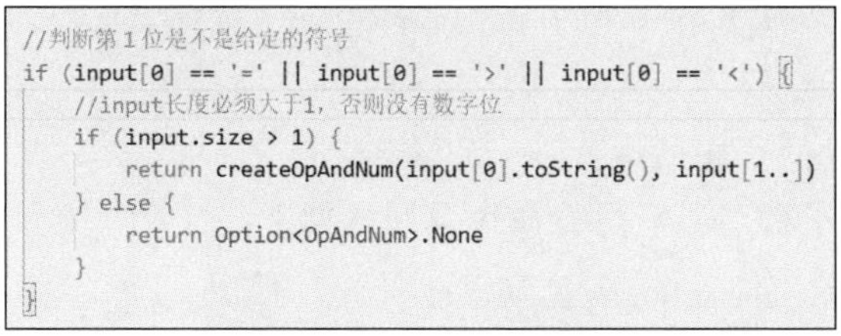

图 2-23　格式化后的代码

# 第3章 第一个仓颉程序

## 3.1 运行 Hello World 程序

做好仓颉的开发准备后,就可以编写第一个仓颉程序了,这里以最简单的 Hello World 程序为例,演示编写、编译及运行的整个过程。

步骤 1:创建 hello_world.cj 的源程序文件,然后使用合适的编辑器进行编辑,代码如下:

```
//Chapter3/hello_world.cj

/*
程序主入口
*/
main() {
    //打印 hello world
    print("Hello, World!\n")
}
```

步骤 2:编译 hello_world.cj 源程序。在 Linux 平台上的编译命令如下:

```
cjc - o hello_world hello_world.cj
```

编译成功后,将会生成 hello_world 可执行文件。

步骤 3:执行 hello_world。命令及回显如下:

```
./hello_world
Hello, World!
```

这样就完成了第 1 个仓颉程序的运行。

如果是 Windows 平台,编译、运行命令及回显如下:

```
cjc - o hello_world.exe hello_world.cj
.\hello_world.exe
Hello, World!
```

## 3.2 仓颉程序基本规则

根据 3.1 节的 Hello World 程序,总结仓颉程序编写的基本规则如下。

**1. 仓颉源程序文件的扩展名是 .cj**

这是仓颉源程序文件的专用扩展名,就像 C 语言的 .c 扩展名、Java 语言的 .java 扩展名一样,是仓颉源程序文件的特有标志,如果把 .cj 改成别的扩展名,则编译时会报错。

**2. 仓颉程序需要有一个程序执行的入口函数 main**

这一点和大部分编程语言类似,程序执行时首先找到 main 函数,从 main 函数开始执行,一个可执行的仓颉程序需要有且只能有一个 main 函数。main 函数的返回值可以是整数类型或者 Unit 类型,main 函数可以没有参数,如果有参数,参数必须是 Array<String> 类型。如果仓颉程序不是可执行的,则可以没有 main 函数。

**3. 仓颉程序对大小写敏感**

不管是仓颉本身的关键字,还是定义的函数、变量名称都对大小写敏感,不能混用。

**4. 仓颉程序源码可以添加注释,注释有规定的语法**

单行注释使用 //,在 // 后面并在换行以前的内容会被编译器忽略;多行注释使用 /* 开始 */ 结束,这两者中间的一行或者多行内容都会被编译器忽略。良好的注释可以增加源程序的可读性,让开发者可以更好地理解程序的逻辑。

**5. 仓颉程序拥有丰富的内置函数库,可以直接使用**

print 函数是仓颉 core 库中的全局函数,除此之外,还有 io 库、net 库、time 库、os 库等几十个内置库,这些库里包含的函数可以在导入所在的包后直接调用,详细的包管理方法会在后续章节介绍。

## 3.3 仓颉程序的编译

仓颉源文件可以使用 cjc 编译命令进行编译,cjc 编译命令功能强大,参数众多,在本节只介绍最简单的几种,更多的参数将在后续章节介绍。

编译命令的格式如下:

```
cjc [options] file(s)
```

其中 options 为可选的参数,file 为一个或者多个待编译的文件,以 3.1 节 hello_world.cj 为例,最简单的编译方式就是直接编译,输入的命令如下:

```
cjc hello_world.cj
```

这样,在 Linux 环境下默认生成的输出文件的名称是 main,在 Windows 环境下是 main.exe。

典型的可选参数如下所示。

1. --output < value >,-o < arg >

设置输出文件的路径,其中-o为简写形式,示例如下:

```
cjc hello_world.cj -- output /data/hello
```

该示例将编译成 hello 文件并输出到/data 目录下。

2. --version,-v

打印编译器版本信息,其中-v为简写形式。

3. --help,-h

显示帮助信息,其中-h为简写形式。

如果一个仓颉程序包含多个源程序文件,例如 a.cj、b.cj、c.cj,则可以在编译时列出这些文件,示例如下:

```
cjc a.cj b.cj c.cj -o d
```

该命令将编译这些源程序文件,并生成 d 文件。

# 第4章 基本数据类型与操作符

## 4.1 标识符与关键字

### 4.1.1 标识符

在仓颉语言中，变量、函数、自定义类型等元素的命名都需要用到标识符，标识符又分为普通标识符和原始标识符，下面分别进行介绍。

**1. 普通标识符**

普通标识符是指除了关键字（见4.1.2节）以外的两种字符序列。

（1）由英文字母开头，后接任意长度的英文字母、数字或下画线(_)。

（2）由一个或多个下画线(_)开头，后接一个或多个英文字母，最后接任意长度的英文字母、数字或下画线(_)。

一些合法的普通标识符示例如下：

```
cangjie
Cang_Jie
cangjie_2024
a
_a
```

一些非法的普通标识符示例如下：

```
_1a                //以下画线(_)开头,但后面没有紧跟着英文字母
2024cangjie        //以数字开头
func               //函数定义的关键字
cang jie           //不是连续的,中间有空格
```

**2. 原始标识符**

原始标识符分为两类。

（1）普通标识符外面加上一对反引号(`)，示例如下：

```
`cangjie`
`Cang_Jie`
`cangjie_2021`
```

(2) 关键字外面加上一对反引号(`),示例如下:

```
`func`
`return`
`let`
```

一些非法的原始标识符示例如下:

```
`_1a`           //以下画线(_)开头,但后面没有紧跟着英文字母
`2021cangjie`   //以数字开头
```

### 4.1.2 关键字

仓颉语言中有一些单词具有特殊用途,被称为关键字,目前一共有 71 个,具体如表 4-1 所示。

表 4-1 关键字

| abstract | as | break | Bool | case | class | catch |
|---|---|---|---|---|---|---|
| continue | Char | do | else | enum | extend | false |
| finally | for | foreign | from | func | Float16 | Float32 |
| Float64 | if | import | in | init | interface | is |
| Int8 | Int16 | Int32 | Int64 | IntNative | let | macro |
| main | match | mut | Nothing | open | operator | override |
| package | private | prop | protected | public | quote | redef |
| return | spawn | static | struct | super | synchronized | this |
| throw | true | try | type | This | unsafe | UInt8 |
| UInt16 | UInt32 | UInt64 | UIntNative | Unit | var | where |
| while | | | | | | |

8min

## 4.2 变量

变量是仓颉语言中一个比较重要的概念,对于没有开发基础的读者来讲,理解起来稍微有点难度,下面从多个角度解释变量的具体含义。

(1) 变量在内存中占有一块存储区域。

(2) 变量有自己的名称,并且变量的名称必须是一个合法的标识符。

(3) 变量有自己的类型,即使定义时没有显式声明类型,也需要保证编译器能正确推断出来。

(4) 变量需要有值,即使初始化时没有赋予初值,也要保证变量在被使用前赋值。

(5) 变量分为不可变变量和可变变量;不可变变量使用 let 关键字定义,初始化后值不

能改变；可变变量使用 var 关键字定义，可以被重复赋值。

### 1. 正确用法演示

下面通过一个示例演示正确的变量用法：

```
//Chapter4/var_demo.cj

main() {
    //定义不可变变量 count
    let count: Int64 = 36

    //定义可变变量 averageScore
    var averageScore: Float64 = 89.5

    //输出两个变量的值
    println("There are ${count} people in the class, with an average score of ${averageScore}.")

    //修改可变变量的值
    averageScore = 96.0

    //重新输出两个变量的值
    println("After the new exam, there are still ${count} people in the class, with an average score of ${averageScore}.")
}
```

编译后运行该示例，输出如下：

```
There are 36 people in the class, with an average score of 89.500000.
After the new exam, there are still 36 people in the class, with an average score of 96.000000.
```

### 2. 错误用法演示

不可变变量的值在初始化后不允许改变，如果修改初始值，则会出现错误，演示代码如下：

```
//Chapter4/error_demo.cj

main() {
    //定义不可变变量 age
    let age:Int64 = 18

    //定义可变变量 height
    var height:Float64 = 1.72

    //输出两个变量的值
    println("Your age is ${age} and height is ${height}")

    //修改可变变量的值
    height = 1.74
    age = 19
```

```
    //重新输出两个变量的值
    println("After a year,Your age is ${age} and height is ${height}")
}
```

在这段代码里,尝试将不可变变量 age 的值修改为 19,这时在代码编辑器里可能会给出错误提示,如果编辑器是 VS Code,则提示信息如图 4-1 所示,在编译时,也会给出编译错误提示,如图 4-2 所示。

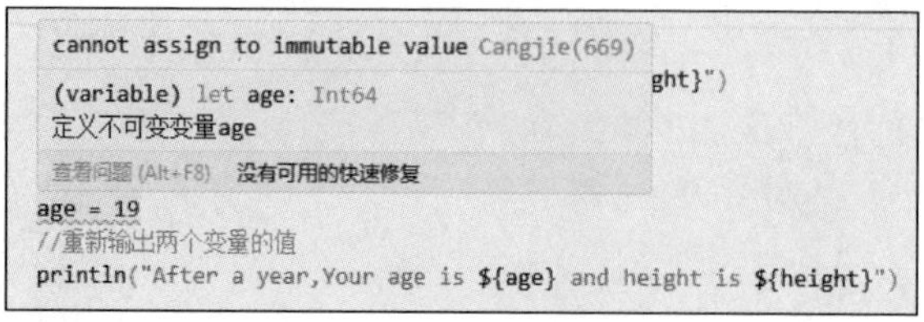

图 4-1　编辑器错误提示

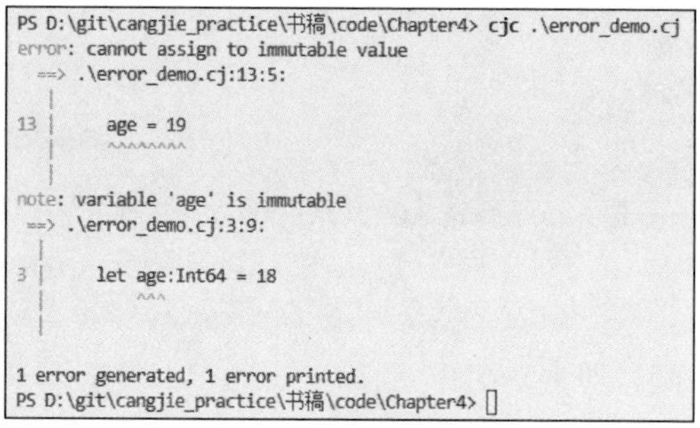

图 4-2　编译器错误提示

### 3. 变量类型推断

如果编译器根据变量的初始值可以推断出变量的类型,则可以在定义变量时省略类型标注,演示代码如下:

```
//Chapter4/deduce_demo.cj

main() {
    //定义不可变变量 count
    let count = 36
    
    //定义可变变量 averageScore
    var averageScore = 89.5
    
    //输出两个变量的值
    println("There are ${count} people in the class, with an average score of ${averageScore}.")
    
    //修改可变变量的值
    averageScore = 96.0
    
    //重新输出两个变量的值
```

```
        println("After the new exam, there are still ${count} people in the class, with an average
          score of ${averageScore}.")
}
```

对这段代码进行编译运行,可以得到和本节第 1 部分"正确用法演示"一样的输出结果。

### 4. 全局变量

定义在源文件顶层的变量被称为全局变量,全局变量在定义时必须初始化,全局变量的作用范围从定义开始到文件结束。

### 5. 局部变量延迟初始化

在函数内部定义的变量被称为局部变量,局部变量在定义时也可以不进行初始化,在这种情况下必须标注类型。仓颉语言要求变量在使用前必须初始化,所以,对于没有在定义时初始化的局部变量,在第 1 次读取变量时,必须确保变量已被初始化,下面通过示例演示局部变量的延迟初始化,代码如下:

```
//Chapter4/delay_init_demo.cj

main() {
    //定义不可变变量 age
    let age: Int64

    //定义可变变量 height
    var height: Float64

    //初始化不可变变量
    age = 18

    //修改可变变量的值
    height = 1.72

    //输出两个变量的值
    println("Your age is ${age} and height is ${height}")
}
```

编译后运行该示例,输出如下:

```
Your age is 18 and height is 1.720000
```

## 4.3 基本数据类型

### 4.3.1 整数类型

#### 1. 类型简介

整数类型分为有符号整数类型和无符号整数类型,在二进制表示形式中,有符号整数类型的最高位代表符号位,0 表示正号(+),1 表示负号(-),其余位表示数值大小;无符号整数类型所有位都用来表示数的大小。以二进制的表示形式 10001010 为例，如果是有符号整数类型，就表示十进制的 −10，如果是无符号整数类型，就表示十进制的 138。

按照有无符号和占用的字节数的不同，整数类型可以分为 10 种更详细的类型，如表 4-2 所示。

表 4-2 整数类型

| 类 型 | 是否有符号 | 占用字节数 | 表 示 范 围 |
|---|---|---|---|
| Int8 | 是 | 1 | $-2^7 \sim 2^7-1(-128 \sim 127)$ |
| Int16 | 是 | 2 | $-2^{15} \sim 2^{15}-1(-32\,768 \sim 32\,767)$ |
| Int32 | 是 | 4 | $-2^{31} \sim 2^{31}-1(-2\,147\,483\,648 \sim 2\,147\,483\,647)$ |
| Int64 | 是 | 8 | $-2^{63} \sim 2^{63}-1(-9\,223\,372\,036\,854\,775\,808 \sim 9\,223\,372\,036\,854\,775\,807)$ |
| IntNative | 是 | 依赖平台 | 依赖平台 |
| UInt8 | 否 | 1 | $0 \sim 2^8-1(0 \sim 255)$ |
| UInt16 | 否 | 2 | $0 \sim 2^{16}-1(0 \sim 65\,535)$ |
| UInt32 | 否 | 4 | $0 \sim 2^{32}-1(0 \sim 4\,294\,967\,295)$ |
| UInt64 | 否 | 8 | $0 \sim 2^{64}-1(0 \sim 18\,446\,744\,073\,709\,551\,615)$ |
| UIntNative | 否 | 依赖平台 | 依赖平台 |

**2. 整数类型字面量的表示形式**

整数类型字面量有 4 种表示形式，如表 4-3 所示，为了方便识别数值位数，可以使用下画线进行分隔。

表 4-3 整数字面量表示形式

| 表 示 形 式 | 前 缀 | 十进制数值 18 表示的示例 |
|---|---|---|
| 二进制 | 0b 或 0B | 0b0001_0010 |
| 八进制 | 0o 或 0O | 0o22 |
| 十进制 | 默认形式，无前缀 | 18 |
| 十六进制 | 0x 或 0X | 0x12 |

下面使用一个简单的示例演示整数类型字面量不同的表示形式，代码如下：

```
//Chapter4/int_demo.cj

func main() {
    //二进制字面量
    let bInt: Int64 = 0b0001_0010

    //八进制字面量
    let oInt: Int64 = 0o22

    //十进制字面量
    let dInt: Int64 = 18

    //十六进制字面量
    let hInt: Int64 = 0x12
```

```
    println("bInt is ${bInt}")
    println("oInt is ${oInt}")
    println("dInt is ${dInt}")
    println("hInt is ${hInt}")
}
```

编译后运行该示例,输出如下:

```
bInt is 18
oInt is 18
dInt is 18
hInt is 18
```

可以看到,这些不同表示形式的字面量实际都是同一个数值。

### 3. 整数类型字面量的推断类型

根据 4.2 节"变量"的说明,变量在有初始值时允许省略定义时的变量标注,这会带来一些困扰,例如有下面的一行变量定义代码:

```
let deduceVar = 10
```

在这行代码里,定义了一个有初始值的整型变量,这个定义还省略了类型标注,因为整型有 10 种不同的具体类型,这个变量的定义根据上下文是没有办法确认具体的类型的,因为 Int64 类型是有符号的,并且可以表示最大的范围,所以,编译器默认会把没有明确标示类型的整型字面量推断为 Int64 类型,这样可以简化代码的编写,也可以统一类型的推断。

### 4. 整数类型字面量的后缀表示

仓颉也支持通过加入后缀的形式来明确整型字面量的类型,例如,要定义一个表示 8 的无符号 8 位整型变量,可以这样实现:

```
let uInt8Var = 8u8
```

后缀和类型的对应关系如表 4-4 所示。

表 4-4 后缀类型对应关系表

| 类 型 | Int8 | Int16 | Int32 | Int64 |
|---|---|---|---|---|
| 后 缀 | i8 | i16 | i32 | i64 |
| 类 型 | UInt8 | UInt16 | UInt32 | UInt64 |
| 后 缀 | u8 | u16 | u32 | u64 |

### 5. 字符字节字面量

UInt8 类型经常会被用来做 ASCII 码字符相关的处理,为了方便使用 UInt8 类型表示字符的 ASCII 码,仓颉引入了字符字节字面量,该字面量由字符 b 和被单引号引用的值组成,示例代码如下:

```
//a 是 UInt8 类型的值 99,也是字符 c 的 ASCII 码值
let a = b'c'
```

```
//b 是 UInt8 类型的值 106,也是字符 j 的 ASCII 码值
let b = b'j'

//c 是 UInt8 类型的值 10,也是换行符的 ASCII 码值
let c = b'\n'
```

### 4.3.2 浮点类型

**1. 类型简介**

浮点类型使用浮点数表示法来表示实数,浮点数表示法是一种科学记数法,目前一般遵循的是 IEEE 制定的 IEEE 754 标准,该表示法使用符号(+或-)、指数和尾数来表示,底数被确定为 2。仓颉中的浮点类型包括 Float16、Float32 和 Float64 共 3 种类型,详细说明如表 4-5 所示。

表 4-5 浮点类型

| 类型 | Float16 | Float32 | Float64 |
| --- | --- | --- | --- |
| 占用字节数 | 2 | 4 | 8 |
| 对应 IEEE 754 格式 | 半精度 binary16 | 单精度 binary32 | 双精度 binary64 |
| 精度 | 小数点后 2 位 | 大概小数点后 6 位 | 大概小数点后 15 位 |
| 符号位 | 1 位 | 1 位 | 1 位 |
| 指数位 | 5 位 | 8 位 | 11 位 |
| 尾数为 | 10 位 | 23 位 | 52 位 |
| 大致表示范围 | 6.104e-5～65504 | 3.4e-38～3.4e+38 | 1.7e-308～1.7e+308 |

**2. 浮点类型的字面量的表示形式**

浮点类型字面量有两种形式,分别是十进制和十六进制。

1) 十进制形式

(1) 至少包含一个整数部分或者一个小数部分。

(2) 如果没有小数部分,则必须包含指数部分。

(3) 指数部分的底数为 10,用前缀 e 或者 E 表示。

2) 十六进制形式

(1) 以 0x 或者 0X 为前缀。

(2) 至少包含一个整数部分或者一个小数部分。

(3) 必须包含指数部分。

(4) 指数部分的底数为 2,用前缀 p 或者 P 表示。

示例代码如下:

```
let a: Float16 = 12.34
let b: Float32 = 10e2
let c: Float64 = 0x2.1p4
```

### 3. 浮点类型字面量的后缀表示

仓颉也支持通过加入后缀的形式来明确浮点字面量的类型,例如,要定义一个表示 0.8 的 Float32 变量,可以这样实现:

```
let f32Var = 0.8f32
```

后缀和类型的对应关系如表 4-6 所示。

表 4-6 后缀类型对应关系表

| 类　　型 | Float16 | Float32 | Float64 |
|---|---|---|---|
| 后　　缀 | f16 | f32 | f64 |

### 4. 浮点类型的计算误差

因为浮点数表示法的固有性质,对于大部分实数来讲,使用浮点数类型无法精确地表示,在这种情况下进行浮点数运算首选使用高精度的浮点类型。下面使用一个示例来演示浮点数的计算误差现象,代码如下:

```
//Chapter4/float_demo.cj

main() {
    //定义一个32位的浮点数,使其字面量为100
    var money: Float32 = 100.0

    //对浮点数减去0.1,此时理论上应该是99.9
    money = money - 0.1

    //输出浮点数的值
    println(money)

    //对浮点数除以3,理论上应该是33.3
    money = money / 3.0

    //输出浮点数的值
    println(money)
}
```

编译后运行该示例,输出如下:

```
99.900002
33.299999
```

可以看到,输出值和理论值有一定的误差,读者可以把 Float32 更换为高精度的 Float64 类型,看一看输出是否会改变。

## 4.3.3 布尔类型

布尔类型表示逻辑关系中的真和假,使用 Bool 关键字进行标识,布尔类型的字面量只有 true 和 false 两个,具体的定义示例如下:

```
let win: Bool = true
var agree = false
```

### 4.3.4 字符类型

字符类型表示 Unicode 字符集中的一个字符,使用关键字 Char 进行标识,它的字面量有 3 种形式,都使用一对单引号定义。

#### 1. 单个字符

直接使用单个字符指定字面量,示例代码如下:

```
let grade: Char = '1'
var sex = 'F'
```

#### 2. 转义字符

使用转义字符表示特殊的字符值,转义字符使用反斜线开头,后跟需要转义的字符,示例代码如下:

```
let tab: Char = '\t'
let newLine = '\n'
```

常用的转义字符如表 4-7 所示。

表 4-7 转义字符

| 转义字符 | 说明 |
| --- | --- |
| \b | 退格,将当前位置移到前一列 |
| \n | 换行,将当前位置移到下一行的开头 |
| \r | 回车,将当前位置移到本行的开头 |
| \t | 制表符,跳到下一个 Tab 位置 |
| \\ | 反斜线 |
| \' | 单引号 |
| \" | 双引号 |

下面使用一段代码,演示部分转义字符的用法,示例代码如下:

```
//Chapter4/escape_demo.cj

main() {
    //打印 Hello 字符串
    print("Hello")
    //打印制表符
    print('\t')
    //打印 World
    print("World")
    //打印换行符
    print('\n')

    //打印 Hello 字符串
```

```
    print("Hello")
    //退格,会把当前位置定位到上一行代码打印出的o的位置
    print('\b')
    //打印World,W将会覆盖Hello的o
    print("World")
    //打印换行符
    print('\n')
}
```

编译后运行该示例,输出如下:

```
Hello          World
HellWorld
```

### 3. 通用字符

使用通用字符表示字面量,通用字符以\u开头,后面加上定义在一对大括号中的1~8个十六进制数,即可表示对应的Unicode值所代表的字符,示例代码如下:

```
//Chapter4/unicode_demo.cj

main() {
    let ch1 = '\u{4f60}'
    let ch2 = '\u{597d}'
    let ch3 = '\u{4ed3}'
    let ch4 = '\u{9889}'
    println(" ${ch1} ${ch2}, ${ch3} ${ch4}")
}
```

编译后运行该示例,输出如下:

```
你好,仓颉
```

## 4.3.5 字符串类型

### 1. 类型简介

单个或者多个Unicode字符组合构成了文本,这种类型就是字符串类型,用关键字String标识。

### 2. 字面量表示形式

字符串的字面量有3种表示形式,分别是单行字符串、多行字符串及多行原始字符串。

1) 单行字符串

单行字符串外侧使用一对引号,内侧是0个、1个或者多个Unicode字符,要特别注意的是,在字符串中表示引号"和反斜线\时,要使用转义字符;单行字符串要求所有字面量都写在同一行上,不允许换行,示例代码如下:

```
let emptyString: String = ""
var escapeString = "\"cangjie\" is the best language!\n"
```

2) 多行字符串

多行字符串以 3 个双引号开头,然后紧接着一个换行,从下一行开始是内容部分,最后以 3 个双引号结尾。内容部分可以是任意数量的 Unicode 字符,可以跨越多行,需要注意的是反斜线需要转义,示例代码如下:

```
//Chapter4/mul_line_string_demo.cj

main() {
    let mulLineString =
    """
    "cangjie" is the best language!
    " and 'don't need Escape.\n
    There is a blank line above this line.
    """
    print(mulLineString)
}
```

编译后运行该示例,输出如下:

```
"cangjie" is the best language!
" and 'don't need Escape.

There is a blank line above this line.
```

3) 多行原始字符串

多行原始字符串以一个或者多个井号(#)加上一个双引号开始,然后是内容部分,最后以双引号加上和开头一样数量的井号结束。内容部分可以是任意数量的 Unicode 字符,也支持换行,和多行字符串不同的是,多行原始字符串不会对转义字符转义,只会保持原样,示例代码如下:

```
//Chapter4/mul_line_raw_string_demo.cj

main() {
    let mulLineString = #"
    "cangjie" is the best language!
    " and 'don't need Escape.\n
    There is a blank line above this line.
    "#
    print(mulLineString)
}
```

编译后运行该示例,输出如下:

```
"cangjie" is the best language!
" and 'don't need Escape.\n
There is a blank line above this line.
```

可以看到,这个示例的字符串内容看似和上一个 mul_line_string_demo.cj 示例一样,但是在输出上就截然不同了。

### 3. 插值字符串

在开发中经常会遇到一种情况,就是字符串中的一部分是不确定的,需要根据程序运行时的状态改变,这时可以使用插值字符串。插值字符串可以在字符串字面量(不能是多行原始字符串)中包括一个或者多个插值表达式,插值表达式需要用大括号"{ }"包起来,并且在前面加上$符号。插值表达式可以包含一个或多个声明或者表达式。在最终计算字符串值时,插值表达式所在的位置会被插值表达式最后一项的值替换,整个插值字符串仍表现为一个字符串,示例代码如下:

```
//Chapter4/interpolation_string_demo.cj

main() {
    let country = "Japan"
    let currency = "Japanese yen"
    let exchangeRate = 17.7
    let money = 100.0

    let outCurrency = " ${currency} is the currency of ${country}"
    println(outCurrency)

    let outMoney = " ${money} yuan can be exchanged for ${ let feeRate = 0.01; money * exchangeRate * (1.0 - feeRate)} ${currency}"
    println(outMoney)
}
```

编译后运行该示例,输出如下:

```
Japanese yen is the currency of Japan
100.000000 yuan can be exchanged for 1752.300000 Japanese yen
```

## 4.3.6 Unit 类型

Unit 类型是一个特殊的类型,它只有一个值,也是它的字面量,即{}。它只支持赋值、判断是否相等这几个操作,不支持别的运算。

## 4.3.7 元组类型

### 1. 类型简介

将多个不同的类型组合在一起的类型叫元组类型,它的定义方式如下:

```
(T1,T2,…,TN)
```

其中,T1 到 TN 是任意数据类型,并且至少是二元或以上,不同类型之间使用逗号(,)分隔。元组的长度是固定的,一旦一个元组实例被定义,它的长度不能再被改变;元组类型还是不可变类型,当一个元组实例被定义后,它的内容便不可被更新。

### 2. 字面量表示形式

元组类型的字面量使用(e1, e2, …, eN)表示,其中 e1 到 eN 是表达式,多个表达式之

间使用逗号分隔,示例如下:

```
let student:(String,Int64,Float64) = ("xiaowang",15,1.72)
```

#### 3. 元素操作

元组支持通过下标访问指定位置的元素,格式为 tuple[index],其中 tuple 为元组,index 为下标,index 是从 0 开始,并且小于元组元素个数的整数,示例代码如下:

```
//Chapter4/tuple_demo.cj
main() {
    //定义元组
    var student: (String, Int64, Float64) = ("xiaowang", 15, 1.72)

    //给元组重新赋值
    student = (student[0], student[1] + 1, student[2] + 0.02)

    //输出各个元素的值
    println("Name: ${student[0]} Age: ${student[1]} Height: ${student[2]}")
}
```

编译后运行该示例,输出如下:

```
Name:xiaowang Age:16 Height:1.740000
```

### 4.3.8 区间类型

区间类型用来表示一个具有固定步长的序列,区间类型的实例包括 3 个值,分别是 start、end 和 step,其中 start 是序列的起始值,end 是终止值,step 表示步长,即序列中两个相邻元素之间的差值。

区间类型字面量有两种形式,分别是左闭右开区间和左闭右闭区间,这里的"闭"指的是"包含"的意思,左闭表示序列包含起始值,右开表示序列不包含终止值,右闭表示序列可以包含终止值。

左闭右开区间的格式为 start..end:step,示例代码如下:

```
let range  = 0..4:1        //包括 0、1、2、3
let range2 = 0..4:2        //包括 0、2
let range3 = 0..5:2        //包括 0、2、4
```

左闭右闭区间的格式为 start..=end:step,示例代码如下:

```
let range  = 1..=4:1       //包括 1、2、3、4
let range2 = 1..=4:2       //包括 1、3
let range3 = 1..=5:2       //包括 1、3、5
```

在区间字面量的定义中,可以省略 step,此时 step 默认等于 1;step 也可以是负数,但是 step 不能是 0。区间也可以不包括任何元素,允许空序列的存在,示例代码如下:

```
let range = 1..= 4              //包括 1、2、3、4
let range2 = 5..0:-2            //包括 5、3、1
let start = 5
let end = 3
let range3 = start..= end:2     //空序列
```

### 4.3.9 Noting 类型

Nothing 是一种特殊的类型,它不包含任何值,是所有类型的子类型。

## 4.4 基本数据类型转换

### 4.4.1 数值类型之间的转换

整数类型和浮点类型统称为数值类型,数值类型之间是可以直接转换的,转换的方式如下:

```
T(e)
```

其中,T 为转换后的数值类型,e 为转换前的数值类型实例,实际转换时要注意浮点型转换为整数型后小数部分丢失的问题,示例代码如下:

```
//Chapter4/conversion_demo.cj

main() {
    //定义 16 位的有符号整数型变量
    let oldInt16Value: Int16 = 10

    //依次转换为有符号 8 位整数型、无符号 64 位整数型、32 位浮点型
    let newInt8Value = Int8(oldInt16Value)
    let newUInt64Value = UInt64(oldInt16Value)
    let newFloat32Value = Float32(oldInt16Value)

    println(newInt8Value)
    println(newUInt64Value)
    println(newFloat32Value)

    //定义 64 位浮点型变量
    let oldFloat64Value: Float64 = 123.45
    //转换为 64 位整数型变量
    let newInt64Value = Int64(oldFloat64Value)
    println(newInt64Value)
}
```

编译后运行该示例,输出如下:

```
10
10
```

```
10.000000
123
```

### 4.4.2　Char 和 UInt32 之间的转换

Char 类型转换为 UInt32 类型使用 UInt32(e)函数,其中 e 为 char 类型的表达式,返回结果为 e 代表的字符在 Unicode 中的代码点(Unicode code point)。UInt32 到 Char 类型的转换使用 Char(num)函数,其中 num 为任意类型的整数,当该整数的值在[0x0000,0xD7FF]或[0xE000,0x10FFFF]中时,返回该整数值代表的代码点对应的字符,示例代码如下:

```
//Chapter4/char_conversion_demo.cj

main() {
    let demoChar = 'a'
    let demoUInt32: UInt32 = 100

    //字符转换为无符号整数
    let conversionUInt32 = UInt32(demoChar)

    //无符号整数转换为字符
    let conversionChar = Char(demoUInt32)

    println("The converted value of ${demoChar} is ${conversionUInt32}")
    println("The converted character of ${demoUInt32} is ${conversionChar}")
}
```

编译后运行该示例,输出如下:

```
The converted value of a is 97
The converted character of 100 is d
```

### 4.4.3　类型判断

当需要判断表达式是否是某个指定类型时,可以通过关键字 is 进行判断,格式如下

```
e is T
```

其中,e 是任意的表达式,T 是任意的类型,如果 e 的运行时类型是 T 的子类型(后续章节讲解自定义类型时会详细讲解子类型,这里可以认为 T 本身就是自己的子类型),则 e is T 的值是 true,否则是 false,示例代码如下:

```
//Chapter4/type_check_demo.cj

main() {
    let int32Value: Int32 = 1
    let int64Value = 1
```

```
    //判断变量类型是否为 Int64 类型
    let typeCheck32 = int32Value is Int64
    let typeCheck64 = int64Value is Int64

    println("Variable int32Value is Int64:${typeCheck32}")
    println("Variable int64Value is Int64:${typeCheck64}")
    println("Variable typeCheck64 is Bool:${typeCheck64 is Bool}")
}
```

编译后运行该示例,输出如下:

```
Variable int32Value is Int64:false
Variable int64Value is Int64:true
Variable typeCheck64 is Bool:true
```

## 4.5 操作符

### 4.5.1 算术操作符

算术操作符主要包括一元负号(-)、加法(+)、减法(-)、乘法(*)、除法(/)、取模(%)、幂运算(**)等,下面分别简述用法。

#### 1. 一元负号

该操作符用来对右侧操作数求负,示例代码如下:

```
let oriValue = 1
let newValue2 = -oriValue
```

一元负号操作符适用于整数类型、浮点类型等数值类型。

#### 2. 加法操作符(+)

加法操作符用来对左右操作数进行相加,要求左右操作数类型相同,适用于整数类型、浮点类型等数值类型。也可以用来做字符串的连接操作,得到一个新的字符串,示例代码如下:

```
//Chapter4/string_plus_demo.cj

main() {
    let leftValue = "Hello"
    let rightValue = "Cangjie"
    let newValue = leftValue + " " + rightValue
    println(newValue)
}
```

编译后运行该示例,输出如下:

```
Hello Cangjie
```

#### 3. 减法操作符(-)

减法操作符使用左侧的操作数减去右侧的操作数,要求左右操作数的类型相同,适用于整数类型、浮点类型等数值类型。

**4. 乘法操作符（*）**

乘法操作符计算左右两个操作数的乘积，要求左右操作数的类型相同，适用于整数类型、浮点类型等数值类型。

**5. 除法操作符（/）**

除法操作符使用左侧的操作数除以右侧的操作数，要求左右操作数的类型相同，适用于整数类型、浮点类型等数值类型。除法操作符有以下几点需要注意。

（1）当操作数是整型时，不允许除数是0，否则将会引起异常。

（2）当操作数是浮点型时，允许除数是0.0，此时得到的值是正无穷大或者负无穷大。

（3）当操作数是整型时，得到的运算结果也是整型。

示例代码如下：

```
//Chapter4/div_demo.cj

main() {
    let leftValue = 10
    let rightValue = 4
    let divValue = leftValue / rightValue
    println(divValue)

    let leftValue2 = 10.0
    var rightValue2 = 4.0
    var divValue2 = leftValue2 / rightValue2
    println(divValue2)

    rightValue2 = 0.0
    divValue2 = leftValue2 / rightValue2
    println(divValue2)
}
```

编译后运行该示例，输出如下：

```
2
2.500000
inf
```

**6. 取模操作符（%）**

取模操作符使用右侧的操作数对左侧的操作数取模，要求左右操作数类型相同并且是整数类型，同时要求右操作数不能是0，否则将会导致编译错误或引起异常，示例代码如下：

```
//Chapter4/modulo_demo.cj

main() {
    var leftValue = 5
    let rightValue = 3

    //右操作数对左操作数取模，右操作数是正数
```

```
    var modulo = leftValue % rightValue
    println(modulo)

    leftValue = -5

    //右操作数对左操作数取模,右操作数是负数
    modulo = leftValue % rightValue
    println(modulo)
}
```

编译后运行该示例,输出如下:

```
2
-2
```

**7. 幂运算符(**)**

幂运算符使用右侧的操作数对左侧的操作数进行幂运算,左操作数只能为 Int64 类型或 Float64 类型,并且需满足以下的要求:

(1) 当左操作数类型为 Int64 时,右操作数只能为 UInt64 类型,表达式的类型为 Int64。

(2) 当左操作数类型为 Float64 时,右操作数只能为 Int64 类型或 Float64 类型,表达式的类型为 Float64。

示例代码如下:

```
//Chapter4/exp_demo.cj

main() {
    let leftValue = 2
    let rightValue = 10u64

    let expValue = leftValue ** rightValue
    println(expValue)

    let floatLeftValue = 2.0
    let floatRightValue = -2.0

    let floatExpValue = floatLeftValue ** floatRightValue
    println(floatExpValue)
}
```

编译后运行该示例,输出如下:

```
1024
0.250000
```

**8. 自增操作符(++)**

自增操作符对给定的变量加 1,适用于整数类型,需要注意以下几点。

(1) 自增操作符是单目操作符,只能跟在操作数的后面。

(2) 自增操作符只能操作使用 var 修饰的变量。

示例代码如下:

```
var value = 10
value++
```

### 9. 自减操作符(--)

自减操作符对给定的变量减1,适用于整数类型,需要注意以下几点。
(1) 自减操作符是单目操作符,只能跟在操作数的后面。
(2) 自减操作符只能操作使用 var 修饰的变量。
示例代码如下:

```
var value = 10
value--
```

## 4.5.2 逻辑操作符

逻辑操作符主要包括逻辑非(!)、逻辑与(&&)和逻辑或(||),下面分别简述其用法。

### 1. 逻辑非操作符(!)

逻辑非操作符是单目操作符,只能在操作数的前面,对操作数进行取非操作,适用于布尔类型,示例代码如下:

```
var opValue = true
let newOpValue = !opValue
```

### 2. 逻辑与操作符(&&)

逻辑与操作符对左右操作数进行逻辑与操作,当左右操作数都是 true 时,其返回值为true,否则返回值为 false,适用于布尔类型,示例代码如下:

```
//Chapter4/logic_and_demo.cj

main() {
    let op1 = true
    let op2 = true
    let op3 = false
    let op4 = false

    println(op1 && op2) //true&&true
    println(op1 && op3) //true&&false
    println(op3 && op1) //false&&true
    println(op3 && op4) //false&&false
}
```

编译后运行该示例,输出如下:

```
true
false
```

```
false
false
```

**3. 逻辑或操作符(||)**

逻辑或操作符对左右操作数进行逻辑或操作,当左右操作数至少有一个是 true 时,其返回值为 true,否则返回值为 false,适用于布尔类型,示例代码如下:

```
//Chapter4/logic_or_demo.cj

main() {
    let op1 = true
    let op2 = true
    let op3 = false
    let op4 = false

    println(op1 || op2)            //true||true
    println(op1 || op3)            //true||false
    println(op3 || op1)            //false||true
    println(op3 || op4)            //false||false
}
```

编译后运行该示例,输出如下:

```
true
true
true
false
```

### 4.5.3 位操作符

位操作符是对整数类型的操作数按位操作的操作符,主要包括按位求反(!)、按位与(&)、按位异或(^)、按位或(|)、按位左移(<<)、按位右移(>>),其中按位与、按位异或和按位或要求左右操作数是相同的整数类型。

在介绍具体的位操作符前,先介绍一种把整数按照二进制格式输出的方法,可以方便后续位操作符的演示,示例代码如下:

```
//Chapter4/binary_out_demo.cj

from std import format.Formatter

main() {
    let opValue: Int8 = 0b01010101
    println(opValue.format("08B"))
}
```

这段代码要注意的有两处,第一处是 from std import format.Formatter,用来导入需要的格式化包,具体的用法在后面章节会讲解,这里按照固定格式使用即可。第二处是

opValue.format("08B"),表示对整型变量格式化输出,格式化字符串为"08B",其中,B表示按照二进制格式输出;8 表示一共输出 8 位,0 表示位数不够 8 位时使用前导 0 补足。编译后运行该示例,输出如下:

```
01010101
```

### 1. 按位求反(!)操作符

按位求反操作符是单目操作符,只能在操作数的前面,对操作数按位进行求反操作,也就是 0 变成 1,1 变成 0,示例代码如下:

```
//Chapter4/bit_not_demo.cj

from std import format.Formatter

main() {
    let opValue: UInt8 = 0b10101010
    let newValue = !opValue

    //分别输出原值和按位取反后的值
    println(opValue.format("08B"))
    println(newValue.format("08B"))
}
```

编译后运行该示例,输出如下:

```
10101010
01010101
```

从输出可以看到每一位都取反了。

### 2. 按位与(&)操作符

按位与操作符对左右操作数按位进行与操作,当两个操作数的对应位都是 1 时,返回值中对应的该位也是 1,否则是 0,示例代码如下:

```
//Chapter4/bit_and_demo.cj

from std import format.Formatter

main() {
    let leftValue: Int16 = 0B0000_1111_0000_1111
    let rightValue: Int16 = 0B0011_0011_0011_0011
    let newValue = leftValue & rightValue

    //分别输出左右操作数及按位"与"后的值
    println(leftValue.format("016B"))
    println(rightValue.format("016B"))
    println(newValue.format("016B"))
}
```

编译后运行该示例,输出如下:

```
0000111100001111
0011001100110011
0000001100000011
```

### 3. 按位异或(^)操作符

按位异或操作符对左右操作数按位进行异或操作,当两个操作数对应位的值不一致时,返回值中对应的该位是1,否则是0,示例代码如下:

```
//Chapter4/bit_xor_demo.cj

from std import format.Formatter

main() {
    let leftValue: Int16 = 0B0000_1111_0000_1111
    let rightValue: Int16 = 0B0011_0011_0011_0011
    let newValue = leftValue ^ rightValue

    //分别输出左右操作数及按位异或后的值
    println(leftValue.format("016B"))
    println(rightValue.format("016B"))
    println(newValue.format("016B"))
}
```

编译后运行该示例,输出如下:

```
0000111100001111
0011001100110011
0011110000111100
```

### 4. 按位或(|)操作符

按位或操作符对左右操作数按位进行或操作,当两个操作数对应位的值至少有一个是1时,返回值中对应的该位是1,否则是0,示例代码如下:

```
//Chapter4/bit_or_demo.cj

from std import format.Formatter

main() {
    let leftValue: Int16 = 0B0000_1111_0000_1111
    let rightValue: Int16 = 0B0011_0011_0011_0011
    let newValue = leftValue | rightValue

    //分别输出左右操作数及按位取或后的值
    println(leftValue.format("016B"))
    println(rightValue.format("016B"))
    println(newValue.format("016B"))
}
```

编译后运行该示例,输出如下:

```
0000111100001111
0011001100110011
0011111100111111
```

### 5. 按位左移(<<)操作符

按位左移操作符用来将左操作数的各二进制位全部向左移若干位，移动的位数由右操作数指定，右操作数必须是非负值，其右边空出的位用 0 填补，高位左移溢出则舍弃该高位。在没有溢出时，左移一位相当于原数值乘以 2，示例代码如下：

```
//Chapter4/shift_left_demo.cj

from std import format.Formatter

main() {
    let leftValue: UInt16 = 0B0000_0000_0000_1111
    let rightValue: UInt16 = 2

    //左操作数向左移 2 位
    let newValue = leftValue << rightValue

    println(leftValue.format("016B"))
    println(newValue.format("016B"))
    println(leftValue)                          //原数值
    println(newValue)                           //向左移 2 位相当于乘以 4

    //左操作数向左移 14 位
    let overFlowValue = leftValue << 14
    println(overFlowValue.format("016B"))       //演示溢出效果
}
```

编译后运行该示例，输出如下：

```
0000000000001111
0000000000111100
15
60
1100000000000000
```

### 6. 按位右移(>>)操作符

按位右移操作符用来将左操作数的各二进制位全部向右移若干位，移动的位数由右操作数指定，右操作数必须是非负值，低位移出(舍弃)，高位的空位补符号位，即正数补 0，负数补 1。向右移一位相当于原数值除以 2 并舍弃余数，示例代码如下：

```
//Chapter4/shift_right_demo.cj

from std import format.Formatter

main() {
    let leftValue: Int16 = 0B0000_0000_0000_1111
```

```
    let rightValue = 2

    //左操作数右移2位
    let newValue = leftValue >> rightValue

    println(leftValue.format("016B"))
    println(newValue.format("016B"))
    println(leftValue)               //原数值
    println(newValue)                //右移2位相当于除以4,不保留余数
}
```

编译后运行该示例,输出如下:

```
0000000000001111
0000000000000011
15
3
```

### 4.5.4 关系操作符

关系操作符用来判断左右操作数之间的关系,包括小于(<)、大于(>)、小于或等于(<=)、大于或等于(>=)、相等(==)、不等(!=),比较的结果是一个布尔值,要求左右操作数是同一种数据类型,下面分别简述其用法。

#### 1. 小于(<)操作符

用来判断左操作数是否小于右操作数,如果小于,则返回值为 true,否则返回值为 false。对于数值型操作数,通过值的大小进行比较;对于字符类型或者字符串类型,通过字符对应的 Unicode 代码值进行比较,示例代码如下:

```
//Chapter4/less_demo.cj

main() {
    println(5 < 3)              //数值型比较
    println('c' < 'j')          //字符比较
    println("hello" < "world")  //字符串比较
}
```

编译后运行该示例,输出如下:

```
false
true
true
```

#### 2. 大于(>)操作符

用来判断左操作数是否大于右操作数,如果大于,则返回值为 true,否则返回值为 false。对于数值型操作数,通过值的大小进行比较;对于字符类型或者字符串类型,通过字符对应的 Unicode 代码值进行比较。

3. **小于或等于(<=)操作符**

用来判断左操作数是否小于或等于右操作数,如果小于或等于,则返回值为 true,否则返回值为 false。对于数值型操作数,通过值的大小进行比较;对于字符类型或者字符串类型,通过字符对应的 Unicode 代码值进行比较。

4. **大于或等于(>=)操作符**

用来判断左操作数是否大于或等于右操作数,如果大于或等于,则返回值为 true,否则返回值为 false。对于数值型操作数,通过值的大小进行比较;对于字符类型或者字符串类型,通过字符对应的 Unicode 代码值进行比较。

5. **相等(==)操作符**

用来判断左操作数是否等于右操作数,如果等于,则返回值为 true,否则返回值为 false。对于数值型操作数,通过值的大小进行比较;对于字符类型或者字符串类型,通过字符对应的 Unicode 代码值进行比较;对于布尔型,通过值进行比较;对于 Unit 类型,也通过值进行比较,示例代码如下:

```
//Chapter4/equal_demo.cj

main() {
    let leftStringValue = "hello"
    let leftBoolValue = true

    println(5 == 3)                         //数值型比较
    println('c' == 'j')                     //字符比较
    println(leftStringValue == "hello")     //字符串比较
    println(leftBoolValue == false)         //布尔类型比较
}
```

编译后运行该示例,输出如下:

```
false
false
true
false
```

6. **不等(!=)操作符**

用来判断左操作数是否不等于右操作数,如果不等于,则返回值为 true,否则返回值为 false。对于数值型操作数,通过值的大小进行比较;对于字符类型或者字符串类型,通过字符对应的 Unicode 代码值进行比较;对于布尔型,通过值进行比较;对于 Unit 类型,也通过值进行比较。

### 4.5.5 赋值操作符

基本的赋值操作符是直接赋值符(=),在此基础上,结合算术操作符和位操作符,延伸出复合赋值操作符,包括+=、-=、*=、/=、%=、**=、<<=、>>=、&=、^=、|=,示

例代码如下：

```
//Chapter4/assign_demo.cj

main() {
    var leftNumValue = 10
    //把变量 leftNumValue 加上 5,然后赋值给自己
    leftNumValue += 5
    println(leftNumValue)

    //把变量 leftNumValue 左移 2 位,然后赋值给自己
    leftNumValue <<= 2
    println(leftNumValue)
}
```

编译后运行该示例，输出如下：

```
15
60
```

### 4.5.6 操作符的优先级

操作符的优先级如表 4-8 所示，上面的优先级最高，从上往下优先级降低。

**表 4-8 操作符的优先级**

| 操作符 | 含义 | 示例 |
| --- | --- | --- |
| @ | 宏调用 | @id |
| . | 成员访问 | expr.id |
| [] | 索引 | expr[expr] |
| () | 函数调用 | expr(expr) |
| ++ | 自增 | var++ |
| -- | 自减 | var-- |
| ? | 问号 | expr?.id |
| ! | 逻辑非 | !expr |
| - | 一元负号 | -expr |
| ** | 幂运算 | expr ** expr |
| * | 乘法 | expr * expr |
| / | 除法 | expr/expr |
| % | 取模 | expr%expr |
| + | 加法 | expr+expr |
| - | 减法 | expr-expr |
| << | 按位左移 | expr << expr |
| >> | 按位右移 | expr >> expr |
| < | 小于 | expr < expr |
| <= | 小于或等于 | expr <= expr |

续表

| 操 作 符 | 含 义 | 示 例 |
|---|---|---|
| > | 大于 | expr > expr |
| >= | 大于或等于 | expr >= expr |
| is | 类型检查 | expr is Type |
| as | 类型转换 | expr as Type |
| == | 判等 | expr == expr |
| != | 判不等 | expr != expr |
| & | 按位与 | expr & expr |
| ^ | 按位异或 | expr ^ expr |
| \| | 按位或 | expr \| expr |
| .. | 区间操作符 | expr..expr |
| ..= | 区间操作符 | expr..=expr |
| && | 逻辑与 | expr && expr |
| \|\| | 逻辑或 | expr \|\| expr |
| ?? | 联合操作符 | expr??expr |
| \|> | 管道操作符 | expr \|> expr |
| ~> | 组合操作符 | expr~>expr |
| = | 赋值 | var = expr |
| **= | 复合赋值 | expr **= expr |
| *= | 复合赋值 | expr *= expr |
| /= | 复合赋值 | expr /= expr |
| %= | 复合赋值 | expr %= expr |
| += | 复合赋值 | expr += expr |
| -= | 复合赋值 | expr -= expr |
| <<= | 复合赋值 | expr <<= expr |
| >>= | 复合赋值 | expr >>= expr |
| &= | 复合赋值 | expr &= expr |
| ^= | 复合赋值 | expr ^= expr |
| \|= | 复合赋值 | expr \|= expr |
| &&= | 复合赋值 | expr &&= expr |
| \|\|= | 复合赋值 | expr \|\|= expr |

# 第 5 章 函 数

函数作为程序逻辑的载体,是仓颉中一个非常重要的概念,本章将讲解函数最基本的一些特性,更复杂的特性会在后续章节介绍。

## 5.1 函数的定义

6min

典型的函数定义,示例代码如下:

```
func add(a: Int64, b: Int64): Int64 {
    return a + b
}
```

从这个示例里可以归纳出函数定义的基本规则如下。
(1) 使用关键字 func 标识函数,表示函数定义的开始。
(2) func 关键字后是函数的名称,名称可以是任意的合法标识符。
(3) 在小括号内是可选的参数列表。
(4) 小括号后是可选的函数返回值类型。
(5) 大括号内是函数体。

上面的定义表明,这是一个名称为 add 的函数,参数列表为两个 Int64 类型的参数 a 和 b,该函数的返回值类型为 Int64 类型,在函数体中把 a 和 b 相加后返回。

**说明:** 从 0.28.4 版本开始,main 被认为是一个关键字,仓颉中的 main 函数被当作一种特殊的函数,定义 main 函数时前面不再需要 func 关键字。

## 5.2 参数及函数调用

10min

### 1. 形参、实参及函数调用

函数的参数列表是可选的,一个函数可以没有参数,也可以拥有 1 个或者多个参数,这里的参数又被称为形式参数,简称形参,在函数调用时,传入的实际参数被称为实参。形参在定义时,可以被认为使用了 let 修饰,也就是说在函数体内不能修改形参的值,否则编译时可能会提示 error：cannot assign to value which is an initialized 'let' constant 的错误信息。函数调用的示例代码如下：

```
//Chapter5/call_demo.cj

func add(param1: Int64,param2: Int64): Int64 {      //这里 param1 和 param2 为函数的形参
    return param1 + param2
}

main() {
    var a = 1
    var b = 2
    let c = add(a, b)                                //这里 a 和 b 为函数的实参
    println(c)
}
```

编译后运行该示例,输出如下：

```
3
```

### 2. 非命名参数及命名参数

形参分为两类,分别是非命名参数和命名参数。非命名参数的定义方式为 p:T,其中,p 是参数名称,T 是参数类型,中间使用冒号分隔。以示例 call_demo.cj 源码中定义的 add 函数为例,该函数使用的就是非命名参数,非命名参数不能设置默认值。

命名参数的定义方式为 p!:T,其中,p 是参数名称,T 是参数类型,中间使用叹号加冒号分隔,命名参数还可以设置默认值,示例代码如下：

```
func add(param1!: Int64, param2!: Int64 = 2): Int64 {
    return param1 + param2
}
```

在调用时,命名参数的实参需要使用 p:e 的形式,其中,p 是命名参数名称,e 是实参。对于有多个命名参数的函数,调用时的传参顺序可以和定义时的参数顺序不一致；如果命名参数设置了默认值,则在调用时该参数可以不传实参,在函数体内会使用该参数的默认值作为实参的值,如果传递了实参,则实参的值将会覆盖默认值。参数列表可以同时包括非命名参数和命名参数,当两者共存时,要先定义非命名参数,后定义命名参数,示例代码如下：

```
//Chapter5/multi_params_demo.cj

func lunch(name: String,drink!: String = "牛奶",fruits!: String = "苹果",
    stapleFood!: String = "米饭",count!: Int64 = 1,takeOut!: Bool = false): String {
    return "Name: ${name} Drink: ${drink} Fruits: ${fruits} StapleFood: ${stapleFood} Count: ${count} TakeOut: ${takeOut}"
}

main() {
```

```
    var menu = lunch("小王",drink: "牛奶",fruits: "苹果",stapleFood: "米饭",count: 1,
takeOut: false)              //传递所有参数的调用示例
    println(menu)

    menu = lunch("小王")      //有默认值的命名参数可以不用传参,使用默认值作为参数值
    println(menu)

    menu = lunch("小张",stapleFood: "面条",drink: "咖啡")
                            //命名参数的传参顺序可以和定义时的顺序不一致
    println(menu)
}
```

该示例包括两个函数,一个是 main 函数,另一个是 lunch 函数,lunch 函数包括 6 个参数,第 1 个是非命名参数,其余的参数均为有默认值的命名参数。在 main 函数里对 lunch 函数进行了 3 次调用,演示了命名参数的使用场景。假如 lunch 全部使用非命名参数,那么每次调用 lunch 函数时都需要把 6 个参数写全,并且要保证顺序一致,这样写起来比较烦琐,而且容易出错;如果使用了命名参数,就不用按照顺序传参,传参时指明了参数名称,不容易出错,更重要的是可以利用默认参数,只需把不是默认值的参数传递过去,提高了开发效率。编译后运行该示例,输出如下:

```
Name:小王 Drink:牛奶 Fruits:苹果 StapleFood:米饭 Count:1 TakeOut:false
Name:小王 Drink:牛奶 Fruits:苹果 StapleFood:米饭 Count:1 TakeOut:false
Name:小张 Drink:咖啡 Fruits:苹果 StapleFood:面条 Count:1 TakeOut:false
```

## 5.3 返回值类型

函数被调用后得到的值的类型是返回值类型,在函数定义时可以显式指定返回值类型,也可以不指定返回值类型,而是由编译器推导确定。在显式定义返回值类型时,要求函数体的类型、函数体中所有 return e 表达式中 e 的类型是返回值类型的子类型。如果定义的返回值类型和实际类型不一致,编译器则会提示错误 type of function body is incompatible with return type,错误示例代码如下:

```
func add(param1: Int64,param2: Int64): Int64 {
    return (param1,param2)
}
```

在未显式指定返回值类型时,编译器会根据函数体的类型及函数体中所有的 return 表达式来共同推导出函数的返回值类型,如下例所示,会推导出返回类型为 Int64 类型:

```
func add(param1: Int64,param2: Int64) {
    return param1 + param2
}
```

如果编译器不能推导出返回值类型,或者推导出的返回值类型不一致(例如多个 return e

表达式中的 e 的类型不一致),编译器则会报错。

## 5.4 函数体

### 1. 函数体的类型

函数体本身也有类型,它的类型是函数体中最后一项的类型,若最后一项为表达式,则函数体的类型是此表达式的类型,若最后一项为变量定义、函数声明或函数体为空,则函数体的类型为 Unit,示例代码如下:

```
func add(param1: Int64,param2: Int64): Int64 {
    param1 + param2
}
```

在这个实例中,最后一项是两个 Int64 类型相加的表达式,这个表达式是 Int64 类型的,也表示函数体是 Int64 类型的,这样就和定义的返回值类型匹配起来了。

### 2. return 表达式

在函数体中,执行到 return 表达式时,会立刻终止函数的执行并返回,返回的类型需要和函数定义的返回值类型一致。return 表达式有两种形式,一种是 return expr,expr 为表达式,此时要求 expr 的类型和函数定义的返回值类型一致;另一种是 return,等价于 return(),因为()是 Unit 的字面量,所以要求函数定义的返回值类型是 Unit。

### 3. 函数局部变量对全局变量的遮盖

定义在源文件顶层的变量是全局变量,定义在函数内部的变量是局部变量,如果函数定义在全局变量的定义后面,则全局变量的作用范围也包括这个函数内部。这时,如果在函数内部定义了一个和全局变量同名的局部变量,局部变量就会对全局变量形成遮盖,也就是从同名局部变量定义开始,到函数的结束,用到这个变量的地方,都会使用局部变量,示例代码如下:

```
//Chapter5/local_var_demo.cj

var gloablVar: Int64 = 1

main() {
    println(gloablVar)
    var gloablVar: Int64 = 5
    println(gloablVar)
}
```

编译后运行该示例,输出如下:

```
1
5
```

从示例可以看出,第 1 次打印时还没有定义局部变量,所以打印出的是全局变量的值,第 2 次打印时,因为已经定义了局部变量,对同名全局变量形成了遮盖,所以打印的是局部变量的值。

## 5.5 嵌套函数(局部函数)

7min

在源文件顶层定义的函数称为全局函数,在函数内部也可以定义函数,称为嵌套函数或者局部函数,局部函数的作用范围从局部函数定义开始到外层函数的结束,示例代码如下:

```
//Chapter5/nested_func_demo.cj

main() {
    //嵌套函数 incNum
    func incNum(num: Int64) {
        return num + 1
    }

    var count: Int64 = 5
    count = incNum(count)
    println(count)
}
```

编译后运行该示例,输出如下:

```
6
```

类似局部变量对全局变量的遮盖,局部函数的变量对外层函数的变量及全局变量也构成遮盖关系,示例代码如下:

```
//Chapter5/local_func_var_demo.cj

//定义全局变量 gloablVar
var gloablVar: Int64 = 1

main() {
    println(gloablVar)
    //局部变量 gloablVar 遮盖了全局变量 gloablVar
    var gloablVar: Int64 = 5

    func localFunc() {
        println(gloablVar)
        //嵌套函数 localFunc 中的变量 gloablVar 遮盖了上层的同名变量
        var gloablVar: Int64 = 15
        println(gloablVar)
    }

    localFunc()
}
```

编译后运行该示例,输出如下:

```
1
5
15
```

在这个示例中,一共有 3 个打印语句,第 1 次打印时,只定义了全局变量,所以打印出 1;第 2 次打印时,已经定义了全局函数内的同名局部变量,打印的是该局部变量,所以打印出 5;第 3 次打印时,在局部函数里也定义了同名变量,形成了对外层变量的遮盖,所以打印出 15。

# 第 6 章 流程控制

在前几章介绍的代码示例中,程序执行的流程都是按顺序执行的,也就是从上到下逐行执行,没有中断、跳转或者循环,这是最基本也是最简单的程序执行逻辑,但是,顺序执行不能解决所有的问题,更多的时候,需要对程序的执行流程进行一些特定的控制,例如,根据某个条件选择执行特定的代码,或者对一段代码重复执行,这时就要用到流程控制,流程控制的实现可以具体分为条件表达式、循环表达式和 match 表达式。

## 6.1 条件表达式

条件表达式使用 if 关键字作为标识,有 3 种表现形式。

**1. 单个 if 的条件表达式**

格式如下:

```
if(expr){
    …//代码块
}
```

其中,expr 为布尔类型的表达式,大括号内是当 expr 为 true 时要执行的代码,示例代码如下:

```
//Chapter6/simple_condition_demo.cj

main() {
    var condition = 1

    //条件表达式,判断 condition 是不是等于 1,这里 condition 等于 1,故执行大括号内的打印语句
    if (condition == 1) {
        println("Condition is 1")
    }

    condition = 2

    //条件表达式,判断 condition 是不是等于 3,这里 condition 等于 2,故不执行大括号内的打印语句
```

```
    if (condition == 3) {
        println("Condition is 3")
    }
}
```

编译后运行该示例,输出如下:

```
Condition is 1
```

### 2. 包含 if 和 else 的条件表达式

格式如下:

```
if(expr){
    …//代码快 1
} else {
    …//代码块 2
}
```

其中,expr 为布尔类型的表达式,if 后大括号内是当 expr 为 true 时要执行的代码,否则就执行 else 后大括号内的代码,示例代码如下:

```
//Chapter6/condition_demo.cj

main() {
    var condition = 2

    if (condition == 1) {
        //条件表达式,判断 condition 是不是等于 1,这里 condition 等于 2,故不执行大括号内的打
        //印语句)
        println("Condition is 1")
    } else {
        //上述 if 条件不成立,故执行 else 后大括号内的代码
        println("Condition is not 1")
    }
}
```

编译后运行该示例,输出如下:

```
Condition is not 1
```

### 3. 使用 else if 的条件表达式

格式如下:

```
if(expr1) {
    …//代码块 1
}
else if(expr2) {
    …//代码块 2
}
…
else {
```

```
    …//代码块
}
```

其中,expr1 为布尔类型的表达式,如果 expr1 为 true 就执行代码块 1,否则再判断 expr2 是否为 true,如果值为 true 就执行代码块 2,否则继续往下判断,中间可以有多个 else if 这样的结构,如果 if 和所有的 else if 后的布尔表达式都不为 true,则执行 else 后大括号内的代码。当然,最后的 else 结构也不是必需的,如果程序不需要,则可以省略。使用多个 else if 结构的示例代码如下:

```
//Chapter6/multi_condition_demo.cj

main() {
    //定义变量 score,表示考试成绩
    let score = 90

    if (score == 100) {
        //如果是 100 分,则打印 Perfect
        println("Perfect")
    } else if (score >= 80) {
        //如果大于或等于 80 分,则打印 Excellent
        println("Excellent")
    } else if (score >= 60) {
        //如果大于或等于 60 分,则打印 Good
        println("Good")
    } else {
        //如果上述条件都不成立,则说明不及格,打印 You should study hard
        println("You should study hard")
    }
}
```

编译后运行该示例,输出如下:

```
Excellent
```

## 6.2 循环表达式

循环表达式用于实现在某个条件成立时重复执行某段代码,有 3 种循环表达式,分别是 for in 表达式、while 表达式和 do while 表达式;除此之外,也可以通过 break 表达式跳出循环或者通过 continue 表达式提前结束循环。

### 1. for in 表达式

for in 表达式用来遍历一个序列,格式如下:

```
for(item in sequence) {
    …//代码块
}
```

其中，for 是 for in 表达式的关键字，for 后的小括号内是循环条件，在循环条件里，sequence 是要遍历的序列，可以是区间或者后续章节要讲解的数组、列表等对象；item 是用来绑定序列中每个元素的变量，序列和变量之间使用关键字 in 分隔；大括号中的代码块是循环体，每次循环都要执行循环体，示例代码如下：

```
//Chapter6/for_in_demo.cj

main() {
    let range = 0..5
    for (i in range) {
        println(i)
    }
}
```

编译后运行该示例，输出如下：

```
0
1
2
3
4
```

默认情况下，循环条件会遍历序列中的每个元素并执行循环体，如果要对序列中的元素进行过滤，也就是只对部分元素执行循环体，则可以在循环条件和循环体之间加上一个条件表达式，格式如下：

```
for(item in sequence where expr) {
    …//代码块
}
```

其中，expr 为布尔表达式，只有满足 expr 为 true 的 item，才会执行后面的循环体，示例代码如下：

```
//Chapter6/for_in_condition_demo.cj

main() {
    let range = 0..5
    for (i in range where i > 2) {
        println(i)
    }
}
```

编译后运行该示例，输出如下：

```
3
4
```

for in 表达式还有一点需要注意，就是绑定序列元素的变量在循环体中是不可修改的，它本身相当于局部变量，作用域从定义开始到循环体结束，也会遮盖外层同名变量。

## 2. while 表达式

while 表达式在循环条件成立时会反复执行循环体中的代码，格式如下：

```
while(expr) {
    …//代码块
}
```

其中，while 为 while 表达式的关键字，后面的小括号内为循环条件，循环条件中 expr 为布尔类型的表达式；大括号内为循环体，只要 expr 为 true，就会循环执行循环体中的代码，示例代码如下：

```
//Chapter6/while_demo.cj

main() {
    var count = 0
    while (count < 3) {
        println(count)
        count++
    }
}
```

编译后运行该示例，输出如下：

```
0
1
2
```

下面详细分析该段代码的执行步骤。

第 1 步，定义了整型变量 count，初始值为 0；

第 2 步，进入 while 循环，因为 count 值为 0，满足 count<3 的条件，进入循环体；

第 3 部，打印 count 的值，打印出的是 0；

第 4 步，执行 count 自增操作，执行后 count 值为 1；

第 5 步，重新判断循环条件，因为 count 值为 1，满足循环条件，再次进入循环体；

第 6 步，打印 count 的值，打印出的是 1；

第 7 步，执行 count 自增操作，执行后 count 值为 2；

第 8 步，重新判断循环条件，因为 count 值为 2，满足循环条件，再次进入循环体；

第 9 步，打印 count 的值，打印出的是 2；

第 10 步，执行 count 自增操作，执行后 count 值为 3；

第 11 步，重新判断循环条件，因为 count 值为 3，不满足循环条件，无法进入循环体，循环结束。

## 3. do while 表达式

while 表达式是先判断循环条件，然后执行循环体，而 do while 则是先执行一遍循环体，然后判断循环条件，格式如下：

```
do {
    …//代码块
} while(expr)
```

其中，do 为 do while 表达式的关键字，后面的大括号内为循环体，大括号后面是 while 关键字，小括号内为循环条件，循环条件中 expr 为布尔类型的表达式。do while 表达式会先执行一遍循环体，然后判断循环条件是否成立，如果成立，就会再执行一遍循环体，然后判断循环条件，如此往复，示例代码如下：

```
//Chapter6/do_while_demo.cj

main() {
    var count = 0
    do {
        println(count)
        count++
    } while (count < 3)
}
```

编译后运行该示例，输出如下：

```
0
1
2
```

### 4. break 表达式

break 表达式出现在循环表达式的循环体内，用于跳出循环并终止当前循环表达式的执行。在下面的示例中，分别使用 for in 循环表达式和 while 循环表达式打印第 1 个 3 次幂大于或等于 100 的数字，示例代码如下：

```
//Chapter6/break_demo.cj

main() {
    //for in 示例
    let range = 1..10
    for (i in range) {
        //首先计算 i 的 3 次幂，然后赋给变量 thirdPower
        let thirdPower = i ** 3
        //如果 thirdPower 大于或等于 100 就打印并跳出
        if (thirdPower >= 100) {
            println(i)
            break
        }
    }

    //while 示例
    var num = 1
    while (true) {
```

```
        let thirdPower = num ** 3
        if (thirdPower >= 100) {
            println(num)
            break
        }
        num++
    }
}
```

编译后运行该示例,输出如下:

```
5
5
```

**5. continue 表达式**

continue 表达式也出现在循环表达式的循环体内,用于提前结束当前循环并开启新一轮循环。接下来演示 continue 表达式,打印 1 到 10 之间的偶数,示例代码如下:

```
//Chapter6/continue_demo.cj
main() {
    let range = 1..10
    for (i in range) {
        //如果 i 除以 2 的余数不是 0,则表明是奇数,直接进行下一个循环
        if (i % 2 != 0) {
            continue
        }
        println(i)
    }
}
```

编译后运行该示例,输出如下:

```
2
4
6
8
```

## 6.3 match 表达式

match 表达式使用模式匹配执行 case 分支中的逻辑,在仓颉语言中具有广泛的应用,使用场景也比较多,本节将只介绍最基本的使用方法,更复杂的功能将在后续章节介绍。match 表达式使用 match 关键字,分为包含待匹配值的 match 表达式及不含待匹配值的 match 表达式。

**1. 包含待匹配值的 match 表达式**

首先看一个根据高中年级输出对应内容的 match 表达式,示例代码如下:

```
//Chapter6/match_value_demo.cj

main() {
    let grade = 2
    match (grade) {
        case 1 => println("You\'re a freshman in high school!")
        case 2 => println("You\'re a sophomore in high school!")
        case 3 => println("You\'re a third grader in high school!")
        case _ => println("You\'re not a high school student!")
    }
}
```

编译后运行该示例,输出如下:

```
You're a sophomore in high school!
```

在这个示例中,使用变量 grade 表示年级,然后在 match 表达式里对其值进行匹配,根据匹配的结果输出打印信息。

包含匹配值的 match 表达式,以关键字 match 开头,后跟一对小括号中的待匹配值(可以是任意表达式),然后是定义在大括号中的一系列 case 分支。每个 case 分支都以关键字 case 开头,后跟该分支对应的模式;模式之后是一个可选的额外匹配条件,表示该分支匹配成功后还要满足该匹配条件;接着是 => 符号,之后就是该分支成功匹配后需要执行的操作,该操作可以是一系列的表达式、变量和函数定义(新定义的变量或函数的作用域从其定义处开始到下一个 case 之前结束)。

match 表达式执行时,会把待匹配值依次与每个 case 中的模式进行匹配,如果匹配成功就执行该模式 => 后面的代码,然后退出 match 表达式的执行;如果匹配不成功则继续与下一个 case 中的模式匹配,直到匹配成功。

match 表达式要求所有匹配项必须是穷尽的,也就是待匹配表达式的所有可能值都应该被考虑到,所以,为确保 match 表达式的穷尽性,一般情况下,会在最后一个 case 分支使用通配符模式 _,它可以匹配任何值。

### 2. 不含待匹配值的 match 表达式

这里把 6.1 节第 3 部分使用条件表达式编写的示例使用不含待匹配值的 match 表达式重新写出来,经修改后的示例代码如下:

```
//Chapter6/match_demo.cj

main() {
    //定义变量 score,表示考试成绩
    let score = 90

    match {
        case score == 100 => println("Perfect")     //如果是 100 分,则打印 Perfect
        case score >= 80  => println("Excellent")   //如果大于或等于 80 分,则打印 Excellent
        case score >= 60  => println("Good")        //如果大于或等于 60 分,则打印 Good
```

```
            case _ => println("You should study hard")   //如果上述条件都不成立,则说明不及格,
                                                         //则打印 You should study hard
    }
}
```

编译后运行该示例,输出如下:

```
Excellent
```

不含待匹配值的 match 表达式也是以关键字 match 开头的,但是后面没有待匹配值,而是直接跟着一对大括号,大括号里是一系列的 case 分支。case 分支也以关键字 case 开头,后面不是模式,而是一个 Bool 表达式;接着是=>符号,之后就是该分支成功匹配后需要执行的操作。

不含待匹配值的 match 表达式执行时会依次判断 case 之后表达式的值,如果值为 true,就执行该分支=>后面的代码,然后退出 match 表达式的执行;如果值不为 true,则继续判断下一个 case 后的表达式的值,直到值为 true 或者执行通配符_后的语句。

# 第 7 章 入门综合实例

前 6 章学习了基本的仓颉语言编程知识,每个知识点都有独立的简单示例代码,为了更好地巩固所学的知识,本章将通过一个稍微复杂一点的实例,演示如何使用仓颉程序解决实际问题。

## 7.1 开发需求

### 7.1.1 斐波那契数列

在数学中,有一个非常有名的数列,叫作斐波那契数列,不但在数学中,它在物理、化学等领域也有广泛的应用,这个数列看起来是这样的:

0、1、1、2、3、5、8、13、21、34 …

具体解释起来,这个数列的第 1 项是 0,第 2 项是 1,随后每一项都是前两项的和,在数学中是这样递推定义的:

$F(0) = 0, F(1) = 1, F(n) = F(n-1) + F(n-2)(n >= 2, n \in N*)$

### 7.1.2 要解决的问题

斐波那契数列有很多特性,其中一个特性是随着数列项数的增加,前一项和后一项的比值越来越接近 0.6180339887,也就是黄金分隔比。这里要做的,就是列出前 20 个斐波那契数列的值,并显示每个值对应的前后项比值。

## 7.2 解决思路

### 7.2.1 问题分析

根据 7.1 节的要求,可以知道最终要完成的工作主要有两个,第 1 个是计算出前 20 个斐波那契数列的值,第 2 个是计算每个值和后一个值的比值。这两个工作有前后依赖关系,只有知道了具体的斐波那契数列的值才能计算前后值的比值,所以这个问题的难点就在如何计算斐波那契数列的值上。第2个计算比值的工作也有需要注意的地方,就是斐波那契数列的值都是整数,而比值是小数,如果将两个整数直接相除,则得到的数还是整数,所以需要有一个把整数转换为浮点数进行除运算的过程,这样可以得到比较精确的小数。

### 7.2.2 递归函数

在第5章"函数"中,学习了函数的定义和基本使用,函数的应用方式很广泛,其中一种典型应用就是递归函数,所谓递归函数,就是一个直接或者间接调用了自身的函数。7.1.1节斐波那契数列的数学定义就很类似于递归函数,所以,计算斐波那契数列值的函数可以考虑这样定义:

```
func fib(n: Int64): Int64 {
  if (n == 0) {
    return 0
  } else if (n == 1) {
    return 1
  } else {
    return fib(n - 1) + fib(n - 2)
  }
}
```

上述代码解析如下:

将获取第n个斐波那契数列数值的函数名称定义为fib,参数为n,参数和函数的返回值的类型都是Int64。在函数体里,首先判断n是不是0,如果是0,就直接返回0,然后判断n是不是1,如果是1就直接返回1;这样就把斐波那契数列前两个值确定下来了。最后就是递归调用,如果n不是0或者1,则通过计算fib(n−1)和fib(n−2)的和就可以获取第n个斐波那契数列的值了。当然,严格来讲,在函数的开始还需要判断n是不是负数,读者可以自行添加。

## 7.3 示例代码

这个问题的编程解决方法很多,也有不使用递归的方法,这里给出的示例代码使用比较容易理解的递归方式,完整的代码如下:

```
//Chapter7/fib_demo.cj

main() {
    let range = 0..20
    for (i in range) {
        //第 i 个斐波那契数列的值
        let fibNum = fib(i)
```

```
            //第 i+1 个斐波那契数列的值
            let nexFibNum = fib(i + 1)

            //第 i 和 i+1 个斐波那契数列值的比值
            let ration = Float64(fibNum) / Float64(nexFibNum)

            println("${i}\t${fibNum}\t${ration}")
        }
    }

/*
   计算第 n 个斐波那契数列的值
*/
func fib(n: Int64): Int64 {
    if (n == 0) {
        return 0
    } else if (n == 1) {
        return 1
    } else {
        return fib(n - 1) + fib(n - 2)
    }
}
```

编译后运行该示例,输出如下:

```
0     0       0.000000
1     1       1.000000
2     1       0.500000
3     2       0.666667
4     3       0.600000
5     5       0.625000
6     8       0.615385
7     13      0.619048
8     21      0.617647
9     34      0.618182
10    55      0.617978
11    89      0.618056
12    144     0.618026
13    233     0.618037
14    377     0.618033
15    610     0.618034
16    987     0.618034
17    1597    0.618034
18    2584    0.618034
19    4181    0.618034
```

# 进阶篇

# 第 8 章 struct 类型

## 8.1 长方体引发的思考

假如有一款 3D 处理的软件,其中有这样一种需求,给定两个长方体,要分别计算两个长方体的体积、表面积并比较大小,那么按照前面章节学习的知识,则可能的代码如下:

```
//Chapter8/cuboid_demo.cj

main() {
    //第 1 个长方体的长
    let length1: Float64 = 10.1

    //第 1 个长方体的宽
    let width1: Float64 = 5.5

    //第 1 个长方体的高
    let height1: Float64 = 15.0

    //第 2 个长方体的长
    let length2: Float64 = 8.1

    //第 2 个长方体的宽
    let width2: Float64 = 6.2

    //第 2 个长方体的高
    let height2: Float64 = 13.9

    //第 1 个长方体的表面积
    let area1 = computeArea(length1, width1, height1)

    //第 2 个长方体的表面积
    let area2 = computeArea(length2, width2, height2)

    //打印并比较大小
```

```
    println("The surface area of the first cuboid is ${area1},the second cuboid is ${area2}.")
    if (area1 > area2) {
        println("The first is greater than the second.")
    } else if (area1 < area2) {
        println("The first is less than the second.")
    } else {
        println("The first is equal to the second.")
    }

    //第 1 个长方体的体积
    let volume1 = computeVolume(length1, width1, height1)

    //第 2 个长方体的体积
    let volume2 = computeVolume(length2, width2, height2)

    //打印并比较大小
    println("The volume of the first cuboid is ${volume1},the second cuboid is ${volume2}.")
    if (volume1 > volume2) {
        println("The first is greater than the second.")
    } else if (volume1 < volume2) {
        println("The first is less than the second.")
    } else {
        println("The first is equal to the second.")
    }
}

/*
根据长、宽、高计算长方体的表面积
*/
func computeArea(length: Float64, width: Float64, height: Float64): Float64 {
    return length * width * 2.0 + length * height * 2.0 + width * height * 2.0
}

/*
根据长、宽和高计算长方体的体积
*/
func computeVolume(length: Float64, width: Float64, height: Float64): Float64 {
    return length * width * height
}
```

编译后运行该示例,输出如下:

```
The surface area of the first cuboid is 579.100000,the second cuboid is 497.980000.
The first is greater than the second.
The volume of the first cuboid is 833.250000,the second cuboid is 698.058000.
The first is greater than the second.
```

这段代码能满足计算和比较的要求,只是感觉比较烦琐,在计算表面积和体积时,需要输入的参数较多,如果再计算其他的项目,例如棱长,也需要全部输入这 3 个参数。如果有一种方式,能把这 3 个参数组合到一起,则操作起来会更方便。另外一点,这里计算表面积、体积采用的都是独立的函数,如果还要计算球体、三棱体、十二面体等各种类型的几何体,则会出现大量名称和功能都类似的函数,对这些函数的管理和使用也是比较困难的事情,如果能把相关的函数都集中起来进行管理,并且不相关的函数互相隔离,则可以极大地简化这个工作。这种情况其实是一种比较普遍的要求,仓颉有两种自定义类型能满足这种要求,一种是 struct(结构体),另一种是 class(类),本章将先介绍 struct,后续章节再介绍 class。

## 8.2　struct 类型的定义

　　struct 类型将不同类型的值组合到一起,成为一种新的类型,以 8.1 节的需求为例,对于长方体,可以将类型定义如下:

```
//Chapter8/cuboid_struct_demo.cj

struct Cuboid {
    //长度,成员变量
    let length: Float64

    //宽度,成员变量
    let width: Float64

    //高度,成员变量
    let height: Float64

    /*
      构造函数
      length: 长
      width: 宽
      height: 高
     */
    init(length: Float64, width: Float64, height: Float64) {
        this.length = length
        this.width = width
        this.height = height
    }

    //计算表面积并返回,成员函数
    func area(): Float64 {
        return length * width * 2.0 + length * height * 2.0 + width * height * 2.0
    }

    //计算体积并返回,成员函数
    func volume(): Float64 {
        return length * width * height
    }
}
```

从上述长方体的定义示例可以归纳出 struct 类型的定义：

以关键字 struct 开头，后跟 struct 的名字，接着是定义在一对大括号中的 struct 定义体。struct 定义体中可以定义一系列的成员变量、成员属性（本例没有定义属性，具体参见8.9 节"成员属性"）、构造函数和成员函数。

在上述的示例中，长方体的 struct 类型的名称为 Cuboid，成员变量为 length、width、height，构造函数为 init，成员函数为 area 和 volume。

## 8.3 成员变量

struct 的成员变量的定义方式和 4.2 节"变量"的定义方式一样，使用 let 定义不可变成员变量，使用 var 定义可变成员变量。成员变量还可以分为实例成员变量和静态成员变量，实例成员变量在定义时可以设置初值也可以不设置初值，不设置初值时必须标注类型；静态成员变量要求使用 static 关键字修饰，并且必须有初值，示例代码如下：

```
struct Circle {
    var radius: Float64 = 2.0
    struct let PI: Float64 = 3.14
}
```

在对成员变量的访问中，实例成员变量只能通过实例访问，静态成员变量只能通过 struct 类型的名称访问。假如在程序中创建了上文定义的 Circle 的实例 demoCircle（具体创建方法可参考 8.4 节的"构造函数"），则对成员变量的访问方式如下：

```
//访问实例变量
demoCircle.radius
//访问静态变量
Circle.PI
```

如果直接通过 struct 类型的名称访问 radius，会出现这样的错误提示：'radius' is non-static member, cannot access by type name；如果直接通过实例访问 PI，则会出现这样的错误提示：object cannot access static member 'PI'。

在定义 struct 类型时还需要注意的是不能递归定义或者互递归定义，也就是说，成员变量的类型不能是自身，也不能和其他 struct 类型构成互递归，示例代码如下：

```
//递归定义
struct Recursion {
    let member: Recursion
}

//Recursion1 和 Recursion2 构成互递归定义
struct Recursion1 {
    let member: Recursion2
}
```

```
struct Recursion2 {
    let member: Recursion1
}
```

**说明**：说 a 是 T 类型的实例，指的是 a 是一个 T 类型的值。

## 8.4 构造函数

构造函数负责在创建实例时对所有未初始化的实例成员变量进行初始化，可以分为两类构造函数，一类是普通构造函数，另一类是主构造函数。

### 8.4.1 普通构造函数

普通构造函数以关键字 init 命名，后跟参数列表和函数体，需要注意的是，构造函数不需要标注 func 关键字，也没有返回值类型。在一个 struct 中可以定义多个普通构造函数，需要确保每个普通构造函数和主构造函数都有不同的参数个数，或者参数个数相同，但是参数类型不同。在定义构造函数时，可能会出现函数的参数名称与成员变量重名，这时可以在函数体内使用 this 来区分，this 代表当前实例，使用 this 标注的是成员变量，示例代码如下：

```
//Chapter8/student_demo.cj
struct Student {
    let age: Int64
    var grade = 1
    let name: String
    var sex = "男"

    init(age: Int64, name: String) {
        this.age = age
        this.name = name
    }

    init(age: Int64, name: String, sex: String) {
        this.age = age
        this.name = name
        this.sex = sex
    }

    init(age: Int64, name: String, sex: String, grade: Int64) {
        this.age = age
        this.name = name
        this.sex = sex
        this.grade = grade
    }
}
```

### 8.4.2 主构造函数

一个 struct 类型最多只能有一个主构造函数,主构造函数的名称和 struct 类型的名称相同,示例代码如下:

```
struct Rectangle {
    let width: Int64
    let height: Int64

    Rectangle(width: Int64, height: Int64) {
        this.width = width
        this.height = height
    }
}
```

在 struct 类型的设计中,一种经常出现的情形是定义了没有初始化的成员变量,然后在全参数的主构造函数里对这些变量初始化,那么有没有一种办法简化这种操作呢?答案是有的,主构造函数提供了一种方式,在参数的前面加上 let 或 var,可以将这些参数声明为成员变量,函数体内的成员变量初始化也可以省略,按照这种方式,前面的 Rectangle 定义的简化形式如下:

```
struct Rectangle {
  public Rectangle(let width: Int64, let height: Int64) {}
}
```

当然,也可以在参数里同时使用普通形参和成员变量形参,这时需要把普通形参写在前面,示例代码如下:

```
struct Rectangle {
    let color: String
    public Rectangle(color: String, let width: Int64, let height: Int64) {
        this.color = color
    }
}
```

### 8.4.3 自动生成的无参构造函数

在定义 struct 类型时,如果实例成员变量都有初始值,可以不提供任何构造函数,这时,编译器会自动为该类型生成一个无参构造函数(调用该构造参数时,会创建一个所有实例成员变量的值均等于其初始值的对象),下面是具体的示例,注释部分是自动生成的无参构造函数:

```
struct Rectangle {
  let width: Int64
  let height: Int64
  /* 自动生成的无参构造函数
  public init() {
  }
```

```
    */
}
```

需要注意的是,只要存在一个自定义的构造函数,就不会自动生成无参构造函数。

## 8.5 成员函数

成员函数分为实例成员函数和静态成员函数,实例成员函数只能通过 struct 类型的实例访问,静态成员函数使用 static 关键字修饰,只能通过 struct 类型的名称访问。实例成员函数可以访问静态成员变量,也可以调用静态成员函数;静态成员函数不能访问实例变量,也不可以调用实例成员函数;成员函数对变量、成员函数的访问及调用关系如表 8-1 所示。

表 8-1 访问和调用关系

| 成员函数 | 实例变量 | 静态变量 | 实例成员函数 | 静态成员函数 |
| --- | --- | --- | --- | --- |
| 实例成员函数 | 可以访问 | 可以访问 | 可以调用 | 可以调用 |
| 静态成员函数 | 不可访问 | 可以访问 | 不可调用 | 可以调用 |

成员函数的示例代码如下:

```
struct Circle {
    let radius: Float64
    static let PI: Float64 = 3.14

    public init(radius: Float64) {
        this.radius = radius
    }

    func area() {
        return PI * this.radius * this.radius
    }

    static func getPI() {
        return PI
    }
}
```

## 8.6 可见修饰符

struct 的成员包括成员变量、成员函数及后面要介绍的成员属性,可以使用 public 和 private 两种可见性修饰符修饰;使用 public 修饰的成员,在 struct 定义的内部和外部均可见;使用 private 修饰的成员仅在 struct 定义内部可见,外部无法访问;struct 定义时还可以不写可见性修饰符,这种缺省的成员仅包内可见。struct 可见性修饰符的可见关系如表 8-2 所示。

表 8-2　成员可见性

| 修　饰　符 | 定　义　内 | 包内其他类型 | 包　　外 |
| --- | --- | --- | --- |
| private | 可见 | 不可见 | 不可见 |
| 缺省 | 可见 | 可见 | 不可见 |
| public | 可见 | 可见 | 可见 |

下面通过一个示例演示可见性修饰符的用法,示例代码如下:

```
//Chapter8/visibility_demo.cj
struct Examination {
    //姓名
    public let name: String

    //分数
    private let score: Float64

    //等级
    var grade: Char = 'C'

    Examination(name: String, score: Float64) {
        this.name = name
        this.score = score
        grade = computeGradeByScore(score)
    }

    /*
      根据分数计算等级
     */
    private func computeGradeByScore(score: Float64): Char {
        let result: Char
        if (score >= 90.0) {
            result = 'A'
        } else if (score >= 80.0) {
            result = 'B'
        } else if (score >= 60.0) {
            result = 'C'
        } else {
            result = 'D'
        }
        return result
    }
}
```

在这个示例中,Examination 类型的成员变量 name 使用 public 修饰,是外部可见的,可以被包内或者包外访问;成员变量 grade 的修饰符为缺省状态,仅包内可见;成员变量 score 和成员函数 computeGradeByScore 都是私有的,外部不可见,如果在外部访问,则会出现 can not access field 'score'或者 can not access function 'computeGradeByScore'的错误提示。

## 8.7 实例的创建与访问

struct 类型定义后就可以通过调用构造函数创建实例了,如果实例创建完毕,则可以通过实例访问它的 public 成员,这里还是针对 8.1 节的需求,给出使用 struct 类型的示例,代码如下:

```
//Chapter8/cuboid_struct_compare_demo.cj

struct Cuboid {
    Cuboid(let length: Float64, let width: Float64, let height: Float64) {}

    //表面积
    func area(): Float64 {
        return length * width * 2.0 + length * height * 2.0 + width * height * 2.0
    }

    //体积
    func volume(): Float64 {
        return length * width * height
    }
}

main() {
    //第 1 个长方体
    let cuboid1 = Cuboid(10.1, 5.5, 15.0)

    //第 2 个长方体
    let cuboid2 = Cuboid(8.1, 6.2, 13.9)

    //打印并比较表面积的大小
    println("The surface area of the first cuboid is ${cuboid1.area()},the second cuboid is ${cuboid2.area()}.")
    if (cuboid1.area() > cuboid2.area()) {
        println("The first is greater than the second.")
    } else if (cuboid1.area() < cuboid2.area()) {
        println("The first is less than the second.")
    } else {
        println("The first is equal to the second.")
    }

    //打印并比较体积的大小
    println("The volume of the first cuboid is ${cuboid1.volume()},the second cuboid is ${cuboid2.volume()}.")
    if (cuboid1.volume() > cuboid2.volume()) {
        println("The first is greater than the second.")
    } else if (cuboid1.volume() < cuboid2.volume()) {
        println("The first is less than the second.")
```

```
    } else {
        println("The first is equal to the second.")
    }
}
```

编译后运行该示例,输出如下:

```
The surface area of the first cuboid is 579.100000,the second cuboid is 497.980000.
The first is greater than the second.
The volume of the first cuboid is 833.250000,the second cuboid is 698.058000.
The first is greater than the second.
```

## 8.8　mut 函数

在默认情况下,struct 中的实例成员函数不能修改实例成员变量和实例成员属性,如果一定要修改,则需要在定义实例成员函数时使用关键字 mut 修饰,这样就允许在函数体内修改当前实例了,示例代码如下:

```
//Chapter8/mut_demo.cj
struct Rectangle {
    Rectangle(var length: Int64, var width: Int64) {}
    mut func setWidth(newWidth: Int64) {
        this.width = newWidth
    }
}

main() {
    var rect = Rectangle(2, 1)
    rect.setWidth(3)
    println(rect.width)
}
```

编译后运行该示例,输出如下:

```
3
```

## 8.9　成员属性

假如有一种场景,需要定义一个包括员工姓名和工资的 struct,按照最简单的方式,代码可能如下:

```
struct Employee {
    //姓名
    var name: String
```

```
//工资
var salary: Float64

public init(name: String, salary: Float64) {
    this.name = name
    this.salary = salary
}
}
```

可以根据需要任意设置员工姓名和工资,但是,实际上各个省市都设置了最低工资标准,例如青岛市2022年的最低工资标准是2100元,直接按照这个结构使用,可能出现允许工资低于最低工资标准的情况,那么有没有办法限制工资不低于最低标准呢？也就是说在设置工资时,如果低于2100元就直接设置成2100元。按照上面的结构比较难实现,但是仓颉语言提供了成员属性的机制,可以完美地实现这种控制。

## 8.9.1 属性的定义

按照成员属性的方式,重新实现的 Employee 如下(本段代码是完整代码的一部分):

```
//Chapter8/employee_demo.cj

struct Employee {
    //姓名
    var name: String

    //私有成员变量工资
    private var realSalary: Float64

    public Employee(name: String, realSalary: Float64) {
        this.name = name
        this.realSalary = realSalary
    }

    //对外部公开的成员属性:工资
    mut prop salary: Float64 {
        get() {
            return realSalary
        }
        set(value) {
            if (value < 2100.0) {
                realSalary = 2100.0
            } else {
                realSalary = value
            }
        }
    }
}
```

在这段代码里,演示了典型的成员属性的定义方式：使用关键字 prop 声明属性,后跟具体的属性名称和类型,大括号内是对 getter(属性取值)及 setter(属性赋值)的实现,其中 getter 需要返回属性的定义类型,setter 的参数表示对属性赋值的值。对于使用 mut 修饰的属性,必须同时定义 getter 和 setter;如果属性只需读取,不需要赋值,则不使用 mut 修饰,并且只定义 getter,示例代码如下:

```
//Chapter8/readonly_prop_demo.cj

struct Exam {
    //姓名
    let name: String

    //分数
    private let score: Float64

    //分数等级
    prop grade: Char {
        get() {
            computeGradeByScore(score)
        }
    }

    public Exam(name: String, score: Float64) {
        this.name = name
        this.score = score
    }

    //根据分数计算等级
    private func computeGradeByScore(score: Float64): Char {
        let result: Char
        if (score >= 90.0) {
            result = 'A'
        } else if (score >= 80.0) {
            result = 'B'
        } else if (score >= 60.0) {
            result = 'C'
        } else {
            result = 'D'
        }
        return result
    }
}
```

### 8.9.2 属性的使用

在外部看来,属性和成员变量类似,使用方式也基本一致;对于使用 mut 修饰的属性,可以取值也可以赋值;对于没有使用 mut 修饰的属性,只能取值,不能赋值。属性也分为实例成员属性和静态成员属性,使用方式也和成员变量一样,示例代码如下:

```
//Chapter8/employee_demo.cj

struct Employee {
    //姓名
    var name: String

    //私有成员变量工资
    private var realSalary: Float64

    public Employee(name: String, realSalary: Float64) {
        this.name = name
        this.realSalary = realSalary
    }

    //对外部公开的成员属性:工资
    mut prop salary: Float64 {
        get() {
            return realSalary
        }
        set(value) {
            if (value < 2100.0) {
                realSalary = 2100.0
            } else {
                realSalary = value
            }
        }
    }
}

main() {
    var emp = Employee("张三", .0)
    emp.salary = 1800.0
    println(" ${emp.name}的工资是 ${emp.salary}元")
}
```

编译后运行该示例,输出如下:

```
张三的工资是:2100.000000 元
```

# 第 9 章 class 类型

class 类型是仓颉语言中一个非常重要的概念,也是仓颉面向对象编程的基础,虽然它的外在表现形式和 struct 类似,但是两者有本质的不同,下面将从定义、创建、继承等多方面介绍 class 的使用。

## 9.1 定义

class 类型的定义和 struct 类似,这里使用 class 的方式针对 8.1 节的需求定义长方体,代码如下:

```
//Chapter9/cuboid_class_demo.cj

class Cuboid {
    //长度,成员变量
    let length: Float64

    //宽度,成员变量
    let width: Float64

    //高度,成员变量
    let height: Float64

    //构造函数
    init(length: Float64, width: Float64, height: Float64) {
        this.length = length
        this.width = width
        this.height = height
    }

    //计算表面积并返回,成员函数
    func area(): Float64 {
        return length * width * 2.0 + length * height * 2.0 + width * height * 2.0
    }
```

```
    //计算体积并返回,成员函数
    func volume(): Float64 {
        return length * width * height
    }
}}
```

从上述长方体的示例,可以归纳出 class 类型的定义:

以关键字 class 开头,后跟 class 的名字,接着是定义在一对大括号中的 class 定义体。class 定义体中可以定义一系列的成员变量、成员属性、构造函数和成员函数。

在上述的示例中,长方体的 class 类型的名称为 Cuboid,成员变量为 length、width、height,构造函数为 init,成员函数为 area 和 volume。

## 9.2 成员变量

class 的成员变量的定义方式和 4.2 节"变量"的定义方式一样,使用 let 定义不可变成员变量,使用 var 定义可变成员变量。成员变量还可以分为实例成员变量和静态成员变量,实例成员变量在定义时可以设置初值也可以不设置初值,不设置初值时必须标注类型;静态成员变量要求使用 static 关键字修饰,并且必须有初值,示例代码如下:

```
class circle {
    var radius: Float64 = 1.0
    static let PI: Float64 = 3.14
}
```

在对成员变量的访问中,实例成员变量只能通过实例访问,静态成员变量只能通过 class 类型名称访问。

## 9.3 构造函数

构造函数负责在创建实例时对所有未初始化的实例成员变量进行初始化,可以分为两类构造函数,一类是普通构造函数,另一类是主构造函数。

### 9.3.1 普通构造函数

普通构造函数以关键字 init 命名,后跟参数列表和函数体。在一个 class 中可以定义多个普通构造函数,需要确保每个普通构造函数和主构造函数都有不同的参数个数,或者参数个数相同,但是参数类型不同。在定义构造函数时,可能会出现函数的参数名称与成员变量重名,这时可以在函数体内使用 this 来区分,this 代表当前实例,使用 this 标注的是成员变量,示例代码如下:

```
//Chapter9/student_class_demo.cj
```

```
class Student {
    let age: Int64
    var grade = 1
    let name: String
    var sex = "男"

    init(age: Int64, name: String) {
        this.age = age
        this.name = name
    }

    init(age: Int64, name: String, sex: String) {
        this.age = age
        this.name = name
        this.sex = sex
    }

    init(age: Int64, name: String, sex: String, grade: Int64) {
        this.age = age
        this.name = name
        this.sex = sex
        this.grade = grade
    }
}
```

## 9.3.2 主构造函数

一个 class 类型最多只能有一个主构造函数，主构造函数的名称和 class 类型的名称相同，示例代码如下：

```
class Rectangle {
  let width: Int64
  let height: Int64

  Rectangle(width: Int64, height: Int64) {
    this.width = width
    this.height = height
  }
}
```

主构造函数还支持成员变量形参，即在参数的前面加上 let 或 var，将这些参数声明为成员变量，函数体内的成员变量初始化也可以省略，按照这种方式，前面的 Rectangle 定义的简化形式如下：

```
class Rectangle {
  Rectangle(let width: Int64, let height: Int64) {}
}
```

当然，也可以在参数里同时使用普通形参和成员变量形参，这时需要把普通形参写在前

面,示例代码如下:

```
class Rectangle {
    let color: String
    Rectangle(color: String, let width: Int64, let height: Int64) {
        this.color = color
    }
}
```

### 9.3.3 自动生成的无参构造函数

在定义 class 类型时,如果不提供任何构造函数,并且所有实例成员变量都有初始值,这时,编译器则会自动为该类型生成一个无参构造函数,当调用该构造函数时,会把所有实例成员变量的值置为初始值。

下面是具体的示例,注释部分是自动生成的无参构造函数,代码如下:

```
class Rectangle {
    let width: Int64 = 1
    let height: Int64 = 1

    /*
        自动生成无参构造函数
    init() {}
    */
}
```

## 9.4 成员函数

成员函数分为实例成员函数和静态成员函数,实例成员函数只能通过 class 类型的实例访问,静态成员函数使用 static 关键字修饰,只能通过 class 类型的名称访问。实例成员函数可以访问静态成员变量,也可以调用静态成员函数;静态成员函数不能访问实例变量,也不可以调用实例成员函数;成员函数对变量、成员函数的访问及调用关系如表 9-1 所示。

表 9-1 访问和调用关系

| 成 员 函 数 | 实 例 变 量 | 静 态 变 量 | 实例成员函数 | 静态成员函数 |
| --- | --- | --- | --- | --- |
| 实例成员函数 | 可以访问 | 可以访问 | 可以调用 | 可以调用 |
| 静态成员函数 | 不可访问 | 可以访问 | 不可调用 | 可以调用 |

成员函数的示例代码如下:

```
class Circle {
    let radius: Float64 = 1.0
    static let PI: Float64 = 3.14

    func area() {
        return PI * this.radius * this.radius
```

```
    }
    static func getPI() {
        return PI
    }
}
```

## 9.5 成员属性

class 成员属性的定义和 struct 类似，使用关键字 prop 声明属性，后跟具体的属性名称和类型，大括号内是对 getter（属性取值）及 setter（属性赋值）的实现，其中 getter 需要返回属性的定义类型，setter 的参数表示对属性赋值的值。对于使用 mut 修饰的属性，必须同时定义 getter 和 setter；如果属性只需读取，不需要赋值，就不使用 mut 修饰，并且只定义 getter，示例代码如下：

```
class Employee {
    //姓名
    var name: String = ""

    //私有成员变量工资
    private var realSalary: Float64 = 2100.0

    //对外部公开的成员属性:工资
    mut prop salary: Float64 {
        get() {
            return realSalary
        }
        set(value) {
            if (value < 2100.0) {
                realSalary = 2100.0
            } else {
                realSalary = value
            }
        }
    }
}
```

## 9.6 可见性修饰符

class 的成员包括成员变量、成员属性、成员函数、构造函数，有 3 种可见性修饰符，分别是 public、protected 和 private，分别表示内部及外部均可见、本 class 及子类可见及 class 内部可见。除此之外，如果不使用任何可见性修饰符修饰，则表示可见性是缺省的，这时表示仅包内可见，详细的示例将在 9.9 节"继承"部分演示。

## 9.7 对象

### 9.7.1 对象的创建与访问

6min

class 类型定义后就可以使用构造函数创建对象了,如果对象创建完毕,则可以通过对象访问它的成员,这里还是针对 8.1 节的需求,给出使用 class 类型的示例,代码如下:

```
//Chapter9/cuboid_class_compare_demo.cj

class Cuboid {
    //构造函数
    Cuboid(var length: Float64, var width: Float64, var height: Float64) {}

    //计算表面积并返回,成员函数
    func area(): Float64 {
        return length * width * 2.0 + length * height * 2.0 + width * height * 2.0
    }

    //计算体积并返回,成员函数
    func volume(): Float64 {
        return length * width * height
    }
}

main() {
    //第 1 个长方体
    let cuboid1 = Cuboid(10.1, 5.5, 15.0)

    //第 2 个长方体
    let cuboid2 = Cuboid(8.1, 6.2, 13.9)

    //打印并比较表面积的大小
    println("The surface area of the first cuboid is ${cuboid1.area()},the second cuboid is ${cuboid2.area()}.")
    if (cuboid1.area() > cuboid2.area()) {
        println("The first is greater than the second.")
    } else if (cuboid1.area() < cuboid2.area()) {
        println("The first is less than the second.")
    } else {
        println("The first is equal to the second.")
    }

    //打印并比较体积的大小
    println("The volume of the first cuboid is ${cuboid1.volume()},the second cuboid is ${cuboid2.volume()}.")
    if (cuboid1.volume() > cuboid2.volume()) {
        println("The first is greater than the second.")
```

```
        } else if (cuboid1.volume() < cuboid2.volume()) {
            println("The first is less than the second.")
        } else {
            println("The first is equal to the second.")
        }
    }
```

编译后运行该示例,输出如下:

```
The surface area of the first cuboid is 579.100000, the second cuboid is 497.980000.
The first is greater than the second.
The volume of the first cuboid is 833.250000, the second cuboid is 698.058000.
The first is greater than the second.
```

### 9.7.2 对象值的修改

struct 是值类型的,而 class 是引用类型的,下面通过一个示例演示这两者之间的区别,代码如下:

```
//Chapter9/modify_value_demo.cj

class Rectangle {
    Rectangle(var height: Int64, var width: Int64) {}
}

struct Rect {
    Rect(var height: Int64, var width: Int64) {}
}

main() {
    //定义并分别实例化基于 class 和 struct 的两个变量
    let rectangleOri = Rectangle(1, 2)
    var rectOri = Rect(1, 2)

    //分别把 class 和 struct 变量赋值给另一个变量
    let rectangleCopy = rectangleOri
    var rectCopy = rectOri

    //分别修改新变量的成员变量值
    rectangleCopy.height = 3
    rectCopy.height = 3

    //分别打印 class 和 struct 新变量和原变量的成员变量值,并比较是否发生了变化
    println("rectangleOri.height = ${rectangleOri.height}rectangleCopy.height = ${rectangleCopy.height}")
    println("rectOri.height = ${rectOri.height}rectCopy.height = ${rectCopy.height}")
}
```

编译后运行该示例,输出如下:

```
rectangleOri.height = 3 rectangleCopy.height = 3
rectOri.height = 1 rectCopy.height = 3
```

在这个示例里,成员变量一样,即定义了 class 类型的 Rectangle 和 struct 类型的 Rect,然后使用同样的参数分别对其实例化,生成了 rectangleOri 和 rectOri,然后把这两个变量分别赋给新的变量 rectangleCopy 和 rectCopy,随后修改新变量的 height 成员变量的值,设置为 3,最后打印对比这 4 个变量的 height 成员变量值。

从打印结果可以分析出,对于 class 类型来讲,变量在赋值时不会复制对象,多个变量还是指向同一个对象,通过一个变量去修改对象成员的值,其他变量对应的对象成员值也会改变。关于对象在实例化、赋值、修改成员变量时的变化如图 9-1 所示(内存地址只是示意,不表示实际地址)。

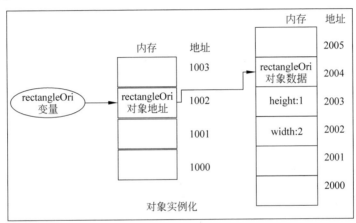

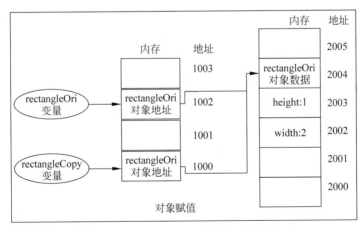

图 9-1　引用类型变量的实例化、赋值及修改

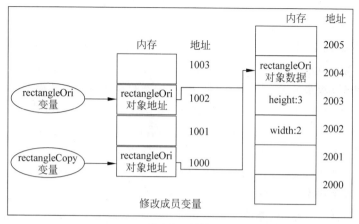

图 9-1 （续）

对于 struct 类型来讲，赋值时会复制一份新的实例给新的变量，这时对新的变量的修改不会影响原来的变量，关于 struct 类型在实例化、赋值、修改成员变量时的变化如图 9-2 所示(内存地址只是示意，不表示实际地址)。

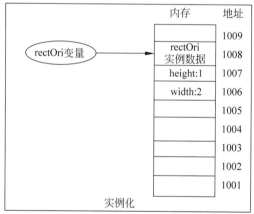

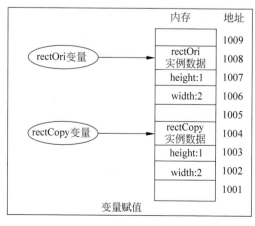

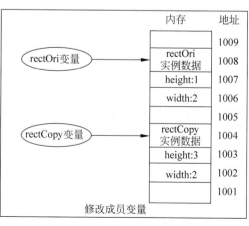

图 9-2 struct 类型变量的实例化、赋值及修改

在作为参数传递时，对象的表现和赋值时一样，此处不再赘述了。在本节的示例中，还需要注意的一点是关于 Rectangle 类型变量 rectangleCopy 的定义，使用的是 let 关键字，在后面修改 rectangleCopy 的 height 成员变量时，程序是可以正常执行的，为什么会这样？这是因为 rectangleCopy 变量存储的是 rectangleOri 的地址，在修改 rectangleCopy 的 height 成员变量时，rectangleCopy 变量存储的地址并没有改变，所以程序可以正常执行。

## 9.8 抽象类

在 class 的成员函数中，有一种函数没有函数体，被称为抽象函数。定义了抽象函数的类被称为抽象类，使用 abstract 修饰，示例代码如下：

```
abstract class Figure {
    public func area(): Float64
    public func perimeter(): Float64
}
```

在这个示例中，定义了表示图形的抽象类，该类包括计算面积和周长的两个抽象函数。抽象函数不能被实例化，就上面的示例来讲，如果调用 Figure 类型的构造函数进行实例化，则会给出错误提示：abstract class 'Figure' can not be instantiated。

## 9.9 继承

### 9.9.1 继承的定义

仓颉语言支持 class 的继承，如果 class B 继承 class A，则称 A 为父类，B 为子类，子类将继承父类中除 private 成员和构造函数以外的所有成员。不是所有的 class 都可以被继承，只有满足下列条件之一的 class 才可以被继承。

（1）抽象类。
（2）定义时使用 open 修饰的类。

仓颉语言的 class 只支持单继承，也就是只能有一个父类，子类在定义时使用<:指定要继承的父类，示例代码如下：

```
open class A {
    let a: Int64 = 1
}

class B <: A {
    let b: Int64 = 1
}

open class C {
    public open func f() {}
```

```
}

class D <: C {
    let d: Int64 = 1
}
```

在继承抽象类时,如果子类不使用 abstract 修饰,则子类必须实现抽象类的所有抽象函数。因为子类继承了父类,子类的对象可以直接作为父类的对象使用,也就是子类类型的对象可以赋值给父类类型的变量,反之则不成立。抽象类继承和子类对象赋值给父类变量的示例代码如下:

```
//Chapter9/abstract_class_demo.cj

abstract class Figure {
    var name: String = "Figure"
    public func area(): Float64
    public func perimeter(): Float64

    func printInfo() {
        println(" The area of the ${name} figure is ${area()} and the perimeter is ${perimeter()}.")
    }
}

class Rectangle <: Figure {
    Rectangle(let height: Float64, let width: Float64) {
        name = "Rectangle"
    }

    public func area() {
        return height * width
    }

    public func perimeter() {
        return (height + width) * 2.0
    }
}

class Circle <: Figure {
    static let PI = 3.14

    Circle(let radius: Float64) {
        name = "Circle"
    }

    public func area() {
        return PI * radius * radius
    }
```

```
    public func perimeter() {
        return 2.0 * PI * radius
    }
}

main() {
    var figure: Figure
    figure = Rectangle(1.0, 2.0)
    figure.printInfo()

    figure = Circle(1.0)
    figure.printInfo()
}
```

编译后运行该示例,输出如下:

```
The area of the Rectangle figure is 2.000000 and the perimeter is 6.000000.
The area of the Circle figure is 3.140000 and the perimeter is 6.280000.
```

在这个示例中,定义了表示图形的抽象类 Figure,它包含计算面积和周长的抽象成员函数 area、perimeter 及成员变量 name,还包含打印图形信息的成员函数 printInfo。抽象类 Figure 实际上描述了图形的一些通用特征,它起到规范子类成员命名的作用,例如计算面积的成员函数都叫 area,计算周长的函数都叫 perimeter,在子类中只需考虑这些函数的实现,而不用再重新命名了。此外,抽象类还可以实现子类中的通用函数,例如 printInfo,这样在各个子类中就不用单独实现了。在上述示例中,表示矩形的 Rectangle 和表示圆形的 Circle 都继承了抽象类 Figure,并且实现了所有的抽象函数。在 main 方法里,定义了 Figure 类型的变量 figure,因为 Figure 是父类,所以两个子类的实例化对象都可以直接赋值给它,并且可以调用它的成员函数。

### 9.9.2 覆盖和重定义

当子类继承父类时,可以实现父类的抽象成员函数,这时可以把子类的业务逻辑放在函数的具体实现中。在某些情况下,父类的普通成员函数可能无法满足子类的要求,这时是不能直接重新实现该函数的,需要在父类的成员函数上使用 open 关键字修饰,然后在子类的同名函数上使用 override 关键字修饰。在调用被覆盖的成员函数时,会根据实际运行时对象的类型进行判断,如果对象类型是父类,就调用父类的成员函数,否则就调用子类成员函数,示例代码如下:

```
//Chapter9/override_demo.cj

open class Parent {
    public open func printInfo() {
        println("This is parent!")
    }
}
```

```
class Child <: Parent {
    public override func printInfo() {
        println("This is child!")
    }
}

main() {
    var p: Parent = Parent()
    p.printInfo()
    p = Child()
    p.printInfo()
}
```

编译后运行该示例,输出如下:

```
This is parent!
This is child!
```

对于父类的非抽象静态成员函数,也可以在子类中定义新的实现,方式为在子类的同名静态成员函数前使用关键字 redef 修饰,示例代码如下:

```
//Chapter9/redef_demo.cj

open class Parent {
    static func printInfo() {
        println("This is parent!")
    }
}

class Child <: Parent {
    static redef func printInfo() {
        println("This is child!")
    }
}

main() {
    Parent.printInfo()
    Child.printInfo()
}
```

编译后运行该示例,输出如下:

```
This is parent!
This is child!
```

### 9.9.3 super 关键字

在子类中,如果覆盖了父类的成员函数,但是还需要访问被覆盖的函数,这时就可以使用 super 关键字了,super 在子类中使用,表示父类对象,可以使用 super 访问父类中的

public 和 protected 成员，包括构造函数，示例代码如下：

```
//Chapter9/super_demo.cj

open class Parent {
    let pMember: Int64

    init(pMember: Int64) {
        this.pMember = pMember
    }

    public open func printInfo() {
        println("This is parent,pMember is ${pMember}")
    }
}

class Child <: Parent {
    let cMember: String

    init(pMember: Int64, cMember: String) {
        super(pMember)
        this.cMember = cMember
    }

    public override func printInfo() {
        super.printInfo()
        println("This is child,cMember is ${cMember}")
    }
}

main() {
    var c: Child = Child(1, "child")
    c.printInfo()
}
```

编译后运行该示例，输出如下：

```
This is parent,pMember is 1
This is child,cMember is child
```

在这个示例中，定义了父类 Parent 及子类 Child，在子类的构造函数中使用 super() 调用了父类的构造函数，在子类的 printInfo 成员函数中，也使用 super 调用了父类的成员函数。

### 9.9.4　成员可见性

在可见性修饰符中，public 比较容易理解，它修饰的成员在类的内部和外部都是可见的，缺省的修饰也和 public 类似，其区别是仅包内可见，重点要关注的是 protected 和 private。

class 可见性修饰符的可见关系如表 9-2 所示。

表 9-2　成员可见性

| 修饰符 | 定义内 | 子类 | 包内其他类型 | 包外 |
| --- | --- | --- | --- | --- |
| private | 可见 | 不可见 | 不可见 | 不可见 |
| protected | 可见 | 可见（包外子类也可见） | 可见 | 不可见 |
| 缺省 | 可见 | 可见（包外子类不可见） | 可见 | 不可见 |
| public | 可见 | 可见 | 可见 | 可见 |

下面通过两个源码文件演示成员的可见性，其中一个文件的名称为 parent.cj，位于 src 目录下的 other 子目录，另一个文件 visibility_demo.cj 位于 src 目录，代码如下：

```
//Chapter9/other/parent.cj

package other

public open class Parent {
    //public 修饰成员
    public var publicMember: String = "public"

    //缺省修饰成员
    var defaultMember: String = "default"

    //protected 修饰成员
    protected var protectedMember: String = "protected"

    //private 修饰成员
    private var privateMember: String = "private"
}

//同一包内的子类访问
public class SamePackageChild <: Parent {
    func print(): Unit {
        //println(super.privateMember)        //private 成员不允许访问
        println(super.protectedMember)         //protected 成员允许访问
        println(super.defaultMember)           //default 成员允许访问
        println(super.publicMember)            //public 成员允许访问
    }
}

//同一包内的其他类型访问
func testVisibleInSamePackage(): Unit {
    let parent = Parent()

    //println(parent.privateMember)           //private 成员不允许访问
    println(parent.protectedMember)            //protected 成员允许访问
    println(parent.defaultMember)              //default 成员允许访问
    println(parent.publicMember)               //public 成员允许访问
}
```

```
//Chapter9/visibility_demo.cj

import other.Parent

class Child <: Parent {
    //包外子类访问
    func print(): Unit {
        //println(super.privateMember)          //private 成员不允许访问
        println(super.protectedMember)          //protected 成员允许访问
        //println(super.defaultMember)          //default 成员不允许访问
        println(super.publicMember)             //public 成员允许访问
    }
}

//包外其他类型访问
func testVisibleInOtherPackage(): Unit {
    let parent = Parent()

    //println(parent.privateMember)             //private 成员不允许访问
    //println(parent.protectedMember)           //protected 成员不允许访问
    //println(parent.defaultMember)             //default 成员不允许访问
    println(parent.publicMember)
}
```

在这个示例中,class 类型的 Parent 是父类,包括使用 public、protected、private 和缺省 4 种访问类型进行修饰的成员,通过包内访问、包外访问、包内子类访问、包外子类访问演示了访问修饰符的详细用法。

# 第 10 章 enum 类型

## 10.1 enum 类型的定义

在软件开发中,如果一种事物可能的取值是有限的,并且便于一一列出,则可以考虑使用仓颉语言中的 enum 类型进行表示。以中学的 3 门主科语文、数学、英语为例,可以定义 enum 类型如下:

```
enum Course {
    | Chinese | Mathematics | English
}
```

在这个示例中,定义了一个 enum 类型的 Course;enum 类型可能的取值又称为构造器,在本例中,Course 的构造器有 3 种,分别是 Chinese、Mathematics 和 English。

enum 类型的定义以关键字 enum 开头,接着是 enum 类型的名称,后跟一对大括号,大括号内是 enum 体,其中定义了若干构造器,多个构造器之间使用|符号分隔,第 1 个构造器前的|符号是可选的。

enum 类型可以定义成员属性、成员函数、静态函数和操作符函数,要求构造器和这些成员之间不能重名,包含静态函数的示例代码如下:

```
enum Course {
    Chinese | Mathematics | English

    static func printAllMember(){
        println("Chinese")
        println("Mathematics")
        println("English")
    }
}
```

## 10.2 enum 类型的值

enum 类型定义后,可以创建该类型的实例,类型实例只能取值定义中的一个构造器,示例代码如下:

```
enum Course {
  Chinese | Mathematics | English
}

  let course1 = Course.Mathematics
  let course2 = Chinese
```

在这个示例中,变量 course1 使用"类型名称.构造器"初始化,course2 直接使用构造器初始化,两种方式都是允许的,在使用第 2 种方式时,注意不能有名称冲突,也就是其他的类型名、变量名、函数名不能叫 Chinese,否则只能通过 Course.Chinese 初始化。

## 10.3 enum 类型的使用

enum 类型经常使用在 match 表达式中,根据值的不同执行对应的 case 分支,示例代码如下:

```
//Chapter10/enum_demo.cj

enum Course {
    Chinese | Mathematics | English
}

main() {
    let nextCourse = Course.Mathematics

    let printCourse = match (nextCourse) {
        case Chinese => "语文"
        case Mathematics => "数学"
        case English => "英语"
    }

    println("下一节课是 ${printCourse}")
}
```

编译后运行该示例,输出如下:

下一节课是数学

在上述示例中,nextCourse 变量表示下一节课,然后在 match 表达式里对 nextCourse 进行匹配,case 后为枚举类型 Course 的常量,当 nextCourse 的构造器和某一个 case 后的模式匹配时,就执行该 case 分支的代码。在本示例中,每个 case 分支执行的结果都是返回一个字符串,然后该字符串会被赋值给 printCourse 变量,最后会输出包含该变量的字符串。

## 10.4 有参构造器

enum 类型除了支持无参数的构造器外,还支持带参数的构造器,示例代码如下:

```
//Chapter10/param_enum_demo.cj

enum Course {
    Chinese(UInt8) | Mathematics(UInt8) | English(UInt8)
}

main() {
    let course1 = Course.Chinese(100)
    let course2 = Course.Mathematics(98)
    let course3 = Course.English(90)
    printCourse(course1)
    printCourse(course2)
    printCourse(course3)
}

func printCourse(course: Course) {
    match (course) {
        case Chinese(c) => println("Chinese: ${c}")
        case Mathematics(m) => println("Mathematics: ${m}")
        case English(e) => println("English: ${e}")
    }
}
```

编译后运行该示例,输出如下:

```
Chinese:100
Mathematics:98
English:90
```

在这个示例中把分数作为参数传递给了课程的构造器,在实例化时,需要给构造器传递实参。要注意有参构造器在 match 表达式里的使用,可以把构造器中的参数提取出来存储在分支定义的局部变量里,在分支后面的代码里可以使用该变量。

enum 类型支持定义多个同名构造器,只要这些构造器参数的个数不同即可(无参构造器可以认为是 0 个参数),示例代码如下:

```
enum Course {
    Chinese | Mathematics | English(Char,Char) | Chinese(UInt8) | Mathematics(UInt8) | English(UInt8)
}
```

# 第 11 章　接　　口

## 11.1　为什么需要接口

以建筑房屋为例,房屋建设方需要和门窗的制作方达成一致,约定门窗的建设标准,例如,房门高度为 220cm,宽度为 90cm,厚度为 20cm,开门方式为向内开,这样,建筑房屋时就可以按照这个标准预留门窗位置,门窗的制作方也可以按照这个标准制造产品,一切都按照约定的标准执行,后期可以很容易地完成门窗的实际安装;如果房屋建设方和门窗制作方有任何一方没有按照标准执行,后期门窗的安装就会遇到很大的困难,甚至无法安装。这个双方约定的标准,可以认为是工作对接的接口。

在软件开发中也存在类似的情况,特别是现代的大型软件,可能是由十几人甚至几百人共同开发的,每个人或者小团队负责其中的一个模块,例如 A 程序员负责开发业务逻辑类,B 程序员负责开发工具类,A 程序员需要工具类中的一个函数返回整型数据,但实际上 B 程序员提供的函数返回的是字符串,这样就会出现一系列的协作问题,假如一开始 A、B 程序员就约定好函数的名称、参数类型、返回值类型,就不容易出现这种错误,而这恰好是仓颉语言中接口的应用范围。

## 11.2　接口的定义

一个标准的接口的定义如下:

```
//Chapter11/interface_demo.cj

interface Figure {
    static prop TYPE_NAME: String
    func area(): Float64
    func perimeter(): Float64
}
```

接口使用关键字 interface 开头,后跟接口的名称,紧接着是一对大括号,大括号里是接口的成员。接口可以拥有 0 个、1 个或多个成员,接口的成员都是抽象的,包括成员函数、操作符重载函数(后续章节会介绍)和成员属性。接口成员可以是实例的,也可以是静态的,默认为被 public 修饰,不能被 private 或者 protected 修饰。接口定义和接口成员都可被 open 修饰符修饰,因为接口默认具有 open 语义,open 修饰符在这里是可选的。

## 11.3 接口的实现

10min

### 11.3.1 接口的通常实现

定义了一个接口以后,如果要为某种类型实现该接口,就要在类型里实现接口的所有成员,这种类型也称为该接口的子类型,类型实现接口要满足以下要求。

(1) 对于成员函数和操作符重载函数,要求实现类型提供的函数实现与接口对应的函数名称相同、参数列表相同、返回类型相同(如果函数的返回值类型是 class 类型,则允许实现函数的返回类型是其子类型)。

(2) 对于成员属性,要求是否被 mut 修饰保持一致,并且属性的类型相同。

以 11.2 节定义的接口为例,首先需要定义类 Rectangle 和 Circle,然后为这两个类实现接口,示例代码如下:

```
//Chapter11/interface_demo.cj

//定义接口 Figure
interface Figure {
    static prop TYPE_NAME: String
    func area(): Float64
    func perimeter(): Float64
}

//定义实现了接口 Figure 的类 Rectangle
class Rectangle <: Figure {
    private static let CLASS_TYPE_NAME = "Rectangle"
    public static prop TYPE_NAME: String {
        get() {
            return CLASS_TYPE_NAME
        }
    }

    Rectangle(let height: Float64, let width: Float64) {}

    public func area() {
        return height * width
    }

    public func perimeter() {
        return (height + width) * 2.0
    }
}
```

```
//定义实现了接口 Figure 的类 Circle
class Circle <: Figure {
    static let PI = 3.14
    private static let CLASS_TYPE_NAME = "Circle"

    public static prop TYPE_NAME: String {
        get() {
            return CLASS_TYPE_NAME
        }
    }

    Circle(let radius: Float64) {}

    public func area() {
        return PI * radius * radius
    }

    public func perimeter() {
        return 2.0 * PI * radius
    }
}
```

在这个示例里,类 Rectangle 和 Circle 都实现了接口 Figure,作为 Figure 的子类型,Rectangle 和 Circle 的实例可以作为 Figure 的实例使用。下面通过具体的示例演示使用接口打印 Rectangle 和 Circle 的面积及周长信息,代码如下:

```
//Chapter11/interface_demo.cj

func printInfo(figure: Figure) {
    println("The area of the figure is ${figure.area()} and the perimeter is ${figure.perimeter()}.")
}

main() {
    var figure: Figure

    figure = Rectangle(1.0, 2.0)
    printInfo(figure)

    figure = Circle(1.0)
    printInfo(figure)
}
```

上面的代码只是完整代码的一部分,接口和实现类的代码在本节前面部分已经展示了,编译后运行完整的示例,输出如下:

```
The area of the figure is 2.000000 and the perimeter is 6.000000.
The area of the figure is 3.140000 and the perimeter is 6.280000.
```

在本示例里，定义了打印图形信息的函数 printInfo，它接收一个 Figure 接口类型的参数，在代码里调用接口的实例成员函数以便获取对象信息，虽然在 main 函数中传递实参时，传递的是两个不同的 class 实例，但是因为它们都是接口 Figure 的子类型，所以可以正常传递进去，并且在函数内部都作为 Figure 类型实例使用。

### 11.3.2 接口的默认实现

仓颉语言的所有类型都可以实现接口，但是，对于 class 类型，仓颉语言还支持接口的默认实现，也就是在接口里直接为成员定义实现，在 class 里可以继承该实现，不用再自己提供，示例代码如下：

```
//Chapter11/default_impl_demo.cj

interface DefaultInterface {
    func defaultFunc(): Unit {
        println("Default Interface")
    }
}

//实现了 DefaultInterface 接口的类，它的 defaultFunc 函数使用的是接口的默认实现
class UseDefaultClass <: DefaultInterface {}

//实现了 DefaultInterface 接口的类，它的 defaultFunc 函数使用的是自己的实现
class ImplInterfaceClass <: DefaultInterface {
    public func defaultFunc(): Unit {
        println("Implement Interface")
    }
}

main() {
    var defaultDemo: DefaultInterface

    defaultDemo = UseDefaultClass()
    defaultDemo.defaultFunc()

    defaultDemo = ImplInterfaceClass()
    defaultDemo.defaultFunc()
}
```

编译后运行示例，输出如下：

```
Default Interface
Implement Interface
```

在这个示例中，定义了一个接口 DefaultInterface，它有两个实现，分别是 UseDefaultClass 和 ImplInterfaceClass，第 1 个类使用了默认的接口实现，第 2 个类自己提供了实现，在对两

个类的实例调用接口函数时,会优先使用类本身的函数实现,当本身没有实现时再调用所继承接口的默认实现。

## 11.4 接口的继承

接口可以继承一个或者多个接口,继承多个接口时使用 & 符号分隔被继承的接口,在接口继承时可以添加新的成员,如果某种类型要实现继承了其他接口的接口,就要实现该接口及其继承的接口的所有成员,示例代码如下:

```
interface Plane {
    func fly(): Unit
}

interface Car {
    func run(): Unit
}

interface PlaneCar <: Plane & Car {
    func park(): Unit
}

class FlyingCar <: PlaneCar {
    public func fly() {
        println("fly")
    }

    public func run() {
        println("run")
    }

    public func park() {
        println("park")
    }
}
```

在上例中,接口 PlaneCar 继承了接口 Plane 和接口 Car,类 FlyingCar 实现了接口 PlaneCar,所以它也间接地实现了接口 Plane 和接口 Car,因此类 FlyingCar 也是接口 Plane 和接口 Car 的子类型,使用的示例代码如下:

```
main() {
    let newFlyingCar = FlyingCar()
    let plane: Plane = newFlyingCar
    let car: Car = newFlyingCar
    let planeCar: PlaneCar = newFlyingCar
}
```

## 11.5 类型的多接口实现

类型实现接口时,允许同时实现多个接口,多个接口之间使用 & 符号分隔,接口之间没有顺序要求,示例代码如下:

```
interface Plane {
    func fly(): Unit
}

interface Car {
    func run(): Unit
}

class PlaneCar <: Plane & Car {
    public func fly() {
        println("fly")
    }

    public func run() {
        println("run")
    }
}
```

在实现多接口时,要注意接口成员默认实现可能带来的问题,假如两个接口都有同样签名的成员,并且都有默认实现,那么实现类型就无法选择哪一个是合适的实现,从而导致两个默认实现都失效,这时实现类型必须提供自己的实现,否则会出现编译错误,示例代码如下:

```
interface Plane {
    func park(): Unit {
        println("Plane")
    }
}

interface Car {
    func park(): Unit {
        println("Car")
    }
}

class PlaneCar <: Plane & Car {}
```

在上述示例中,接口 Plane 和接口 Car 都有成员 park 的默认实现,类 PlaneCar 继承了这两个接口,但是没有提供成员 park 的实现,在编译时会出现如下的错误信息:interface function 'park' must be implemented in 'PlaneCar'。

## 11.6 典型的内置接口

仓颉语言内置了丰富的接口,这里介绍比较简单又经常使用的两个。

### 11.6.1 Any 类型

Any 类型是一个内置的特殊接口,定义如下:

```
interface Any {}
```

它没有任何成员变量,仓颉语言中所有接口都默认继承它,所有非接口类型都默认实现它,因此所有类型都是 Any 类型的子类型,这样,Any 类型就有一些看起来奇怪但实际上合理的特性,例如任何变量都可以赋给 Any 类型的变量,示例代码如下:

```
class demoClass {
    let member: Int64 = 1
}

main() {
    var any: Any = 1
    any = demoClass()
    any = "any"
    any = 3.14
}
```

### 11.6.2 ToString 接口

ToString 接口用来返回实例类型的字符串表示,定义如下:

```
public interface ToString {
    func toString(): String
}
```

该接口包含了 toString 成员函数,可以返回一个字符串,仓颉语言内置的基础类型(如 Bool、Char、Float16、Float32、Float64、Int64、Int32、Int16、Int8、UInt64、UInt32、UInt16、UInt8)均实现了该接口。

# 第 12 章 泛　　型

## 12.1　什么是泛型

先通过一个简单的示例了解为什么需要泛型。在开发中经常会遇到这种情形，需要在一个对象里存储索引序号和内容，然后把这个对象存储到列表里，以后可以根据索引序号从列表查找内容。在设计时，一般通过 class 的两个成员变量来分别表示序号和内容，例如，存储内容为字符串类型的 class 的示例代码如下：

```
class StringItem {
    StringItem(let index: Int64, let content: String) {}
}
```

使用该 class 的示例代码如下：

```
let item = StringItem(2, "Hello")
println("index: ${item.index} content: ${item.content}")
```

除了存储字符串，还可能需要存储整型、字符型、浮点型等，甚至可以直接存储一个其他的类对象，一个包含多个内容存储类型的示例代码如下：

```
//Chapter12/non_generic_demo.cj

class Int64Item {
    Int64Item(let index: Int64, let content: Int64) {}
}

class StringItem {
    StringItem(let index: Int64, let content: String) {}
}

class CharItem {
    CharItem(let index: Int64, let content: Char) {}
}

class Float32Item {
```

```
        Float32Item(let index: Int64, let content: Float32) {}
}

main() {
    let item1 = Int64Item(1, 1)
    println("index: ${item1.index} content: ${item1.content}")

    let item2 = StringItem(2, "Hello")
    println("index: ${item2.index} content: ${item2.content}")

    let item3 = CharItem(3, 'A')
    println("index: ${item3.index} content: ${item3.content}")

    let item4 = Float32Item(4, 8.8)
    println("index: ${item4.index} content: ${item4.content}")
}
```

编译后运行示例,输出如下:

```
index:1 content:1
index:2 content:Hello
index:3 content:A
index:4 content:8.800000
```

这个示例包含了存储 64 位整型、字符串型、字符型、32 位浮点型的自定义 class,看起来有点烦琐,因为仓颉语言支持的数据类型很多,基本数据类型就有十几种,如果要对每种类型写一个存储该类型的 class,工作量会非常大,那么有没有办法简化呢?

分析一下上述示例中的自定义 class,可以发现它们的写法很类似,差别就在数据类型上,如果把数据类型也作为参数传递到类的定义中去,就可以简化程序的编写。仓颉语言的泛型类型可以实现该功能,泛型指的是参数化类型,参数化类型是一个在声明时未知并且需要在使用时指定的类型。使用泛型对上述示例重新改造,改造后的代码如下:

```
//Chapter12/generic_demo.cj

class Item<T> {
    Item(let index: Int64, let content: T) {}
}

main() {
    let item1 = Item<Int64>(1, 1)
    println("index: ${item1.index} content: ${item1.content}")

    let item2 = Item<String>(2, "Hello")
    println("index: ${item2.index} content: ${item2.content}")

    let item3 = Item<Char>(3, 'A')
    println("index: ${item3.index} content: ${item3.content}")
```

```
    let item4 = Item<Float32>(4, 8.8)
    println("index:${item4.index} content:${item4.content}")
}
```

编译后运行示例,输出如下:

```
index:1 content:1
index:2 content:Hello
index:3 content:A
index:4 content:8.800000
```

泛型可以用在类型或者函数的声明中,这些在使用处被指定的类型(可能是一个或者多个)被称为类型形参,类型形参需要给定一个标识符(上例中在尖括号中的 T 便是类型形参),一般在类型名称或者函数名的后面,被一对尖括号括起来,如果具有多种类型形参,类型形参之间使用逗号分隔;声明类型形参后,在声明体中通过标识符对类型形参的引用被称为类型变元(上例中 let content:T 中对 T 的引用就是类型变元);在使用泛型声明的类型或函数时,用来指定泛型类型的参数被称为类型实参(上例中代码行 let item1 = Item<Int64>(1,1) 中的 Int64 就是类型实参);泛型中需要零个、一个或者多种类型作为实参的类型称为类型构造器(上例中 Item 便是类型构造器)。

## 12.2 泛型接口

仓颉语言支持定义泛型接口,示例代码如下:

```
//Chapter12/generic_interface_demo.cj

//定义提供比较函数的接口
interface Comparable<T> {
    func compare(obj: T): Int64
}

//实现了比较接口 Comparable 的类 Rectangle
class Rectangle <: Comparable<Rectangle> {
    Rectangle(let width: Float64, let height: Float64) {}

    //对两个 Rectangle 对象进行比较的函数
    public func compare(obj: Rectangle) {
        if (this.area() > obj.area()) {
            return 1
        } else if (this.area() < obj.area()) {
            return -1
        } else {
            return 0
        }
    }
```

```
    func area() {
        return width * height
    }
}

//从两个 Rectangle 对象中选择较大的那一个
func selectLargerOne(obj1: Rectangle, obj2: Rectangle): Rectangle {
    if (obj1.compare(obj2) >= 0) {
        return obj1
    } else {
        return obj2
    }
}

main() {
    let rect1: Rectangle = Rectangle(2.0, 3.0)
    let rect2: Rectangle = Rectangle(3.0, 4.0)
    let bigOne = selectLargerOne(rect1, rect2)
    println("The larger figure area is ${bigOne.area()}.")
}
```

编译后运行示例,输出如下:

```
The larger figure area is 12.000000.
```

在上述示例中,定义了一个泛型接口 Comparable,它拥有一个 compare 成员函数,该函数用来对给定的类型实例进行比较,比较的结果是一个 Int64 类型的值。Rectangle 类实现了 Comparable 接口,在函数 compare 中可以按照面积比较大小。示例中定义了一个 selectLargerOne 函数,它用于比较两个 Rectangle 对象的大小并选择较大的那一个(如果两个一样大就选择第 1 个)。

## 12.3 泛型函数

仓颉语言支持定义泛型函数,全局函数的示例代码如下:

```
//Chapter12/generic_function_demo.cj

func selectOne<T>(selFirst: Bool, firstObj: T, secondObj: T): T {
    if (selFirst) {
        return firstObj
    } else {
        return secondObj
    }
}

main() {
    //从两个整型中选择一个
```

```
        let selNum = selectOne<Int64>(true, 100, 20)
        println(selNum)

        //从两个字符串中选择一个
        let selString = selectOne<String>(false, "hello", "cangjie")
        println(selString)
    }
```

编译后运行示例,输出如下:

```
100
cangjie
```

在这个示例里,定义了一个泛型函数 selectOne,它具有 3 个参数,根据第 1 个布尔型参数是 true 或者 false 决定返回的是第 2 个还是第 3 个参数。泛型函数定义时,在函数名后使用尖括号声明类型形参,然后可以在函数形参、返回类型及函数体中对这一类型形参进行引用。

仓颉语言可以在 class、struct 与 enum 中定义静态泛型函数,示例代码如下:

```
//Chapter12/static_generic_function_demo.cj
class Tools {
    static func selectOne<T>(selFirst: Bool, firstObj: T, secondObj: T): T {
        if (selFirst) {
            return firstObj
        } else {
            return secondObj
        }
    }
}
main() {
    let selNum = Tools.selectOne<Int64>(true, 100, 20)
    println(selNum)
    let selString = Tools.selectOne<String>(false, "hello", "cangjie")
    println(selString)
}
```

编译后运行示例,输出如下:

```
100
cangjie
```

## 12.4 泛型约束

先看一个非常简单的泛型函数,示例代码如下:

```
func test<T>(param: T) {
    return param
}
```

在这个泛型函数中,因为 T 可能是任何类型,所以对于参数 param,不能调用它的任何成员,除了返回参数本身以外,不能对它做任何其他操作,这样使用起来就很受限制。为了解决这种局限性问题,仓颉语言提供了泛型约束,泛型约束一般分为接口约束和子类型约束,语法为在函数、类型的声明体之前使用 where 关键字进行声明,对于声明的泛型形参 T1 和 T2,可以使用 where T1 <: Interface, T2 <: Type 这样的方式声明泛型约束,同一种类型变元的多个约束可以使用 & 连接,例如 where T1 <: Interface1 & Interface2;在泛型声明体中,类型变元可以直接使用泛型约束对应的接口或者子类型的成员。

在 12.2 节的示例中,定义了 selectLargerOne 函数来选择两个矩形中较大的那一个,如果还需要选择其他图形,就要重新编写对应的函数,假如类型较多,编写的工作量也会比较大。在学习了泛型约束的概念后,可以使用泛型函数结合泛型约束来解决此类问题,改进后的示例代码如下:

```
//Chapter12/generic_constraints_demo.cj
//定义了 compare 成员函数的泛型接口
interface Comparable<T> {
    func compare(obj: T): Int64
}

//定义实现了 Comparable 泛型接口的类 Rectangle
class Rectangle <: Comparable<Rectangle> {
    Rectangle(let width: Float64, let height: Float64) {}

    //Comparable 泛型接口的实现函数
    public func compare(obj: Rectangle) {
        if (this.area() > obj.area()) {
            return 1
        } else if (this.area() < obj.area()) {
            return -1
        } else {
            return 0
        }
    }

    func area() {
        return width * height
    }
}

//定义实现了 Comparable 泛型接口的类 Circle
class Circle <: Comparable<Circle> {
    static let PI = 3.14

    Circle(let radis: Float64) {}

    public func compare(obj: Circle) {
```

```
            if (this.area() > obj.area()) {
                return 1
            } else if (this.area() < obj.area()) {
                return -1
            } else {
                return 0
            }
        }

        func area() {
            return PI * radis * radis
        }
    }

    //泛型函数,要求参数实现泛型接口 Comparable
    func selectLargerOne<T>(obj1: T, obj2: T) where T <: Comparable<T> {
        if (obj1.compare(obj2) >= 0) {
            return obj1
        } else {
            return obj2
        }
    }

    main() {
        //定义两个矩形
        let rect1: Rectangle = Rectangle(2.0, 3.0)
        let rect2: Rectangle = Rectangle(3.0, 4.0)

        //选择较大的矩形
        let bigRectangle = selectLargerOne(rect1, rect2)

        println("The larger figure area is ${bigRectangle.area()}.")

        //定义两个圆形
        let circle1: Circle = Circle(3.0)
        let circle2: Circle = Circle(2.0)

        //选择较大的圆形
        let bigCircle = selectLargerOne(circle1, circle2)

        println("The larger figure area is ${bigCircle.area()}.")
    }
```

编译后运行示例,输出如下:

```
The larger figure area is 12.000000.
The larger figure area is 28.260000.
```

在这个示例中定义了泛型函数 selectLargerOne,泛型形参 T 的约束为 Comparable 泛型接口,这样在声明体中,类型变元就可以调用接口 Comparable 的 compare 成员函数了。

## 12.5 泛型类型

### 12.5.1 泛型 class

在 12.1 节已经演示了泛型 class 的基本用法,本节将演示包括多个泛型形参及泛型约束的泛型 class 的高级用法。在具体的演示以前,先介绍一个以前使用了多次的仓颉语言内置全局函数 println,定义如下:

```
public func println<T>(arg: T): Unit where T <: ToString
```

该函数是一个泛型函数,接收一个实现了 ToString 接口的参数,用来向控制台输出数据,在输出的末尾会换行,输出的数据为对参数 arg 调用 ToString 接口的 toString 成员函数返回的字符串;任何实现了 ToString 接口的类型都可以通过函数 println 输出。

泛型 class 的高级用法,示例代码如下:

```
//Chapter12/multi_param_generic_class_demo.cj
class DictionaryItem<K, V><: ToString where K <: ToString, V <: ToString {
    private let key: K
    private var value: V

    prop Key: K {
        get() {
            return key
        }
    }

    mut prop Value: V {
        get() {
            return value
        }
        set(newValue) {
            value = newValue
        }
    }

    init(key: K, value: V) {
        this.key = key
        this.value = value
    }

    //实现了 ToString 接口
    public func toString() {
        return "Key: ${key} Value: ${value}"
    }
}
```

```
main() {
    let item1 = DictionaryItem<Int8, Int64>(1, 1)
    println(item1)

    let item2 = DictionaryItem<Char, String>('2', "Hello")
    println(item2)

    let item3 = DictionaryItem<Int64, Char>(3, 'A')
    println(item3)

    let item4 = DictionaryItem<String, Float32>("4", 8.8)
    println(item4)
}
```

编译后运行示例,输出如下:

```
Key:1 Value:1
Key:2 Value:Hello
Key:3 Value:A
Key:4 Value:8.800000
```

在这个示例中,定义了一个泛型类 DictionaryItem,它具有两种类型形参,两种类型形参都具有 ToString 的接口约束,同时,该泛型类本身也实现了 ToString 接口。

### 12.5.2 泛型 struct

struct 类型的泛型类似 class 类型的泛型,示例代码如下:

```
//Chapter12/generic_struct_demo.cj

struct RepeatPrint<T> where T <: ToString {
    RepeatPrint(let count: Int64, let content: T) {}

    func repeatPrint() {
        for (i in 0..count) {
            println(content)
        }
    }
}

main() {
    let repeatPrint = RepeatPrint<String>(3, "Hello cangjie!")
    repeatPrint.repeatPrint()
}
```

编译后运行示例,输出如下:

```
Hello cangjie!
Hello cangjie!
Hello cangjie!
```

该示例定义了一个 struct 类型的泛型 RepeatPrint，它接收泛型类型的内容参数，可以把内容重复输出给定的次数。

### 12.5.3 泛型 enum

在泛型 enum 的应用中，最重要的内置类型是 Option 类型，下面通过一个示例解释为什么需要 Option 类型。假设要写一个除法的函数，示例代码如下：

```
//Chapter12/dived_by_zero_demo.cj

func divFunc(dividend: Int64, divisor: Int64): Int64 {
    return dividend / divisor
}

main() {
    let value = divFunc(100, 0)
    println(value)
}
```

编译后运行示例，输出如下：

```
An exception has occurred:
ArithmeticException: Divided by zero !
        at rt$ThrowArithmeticException(std\core\runtime_call_throw_exception.cj:19)
        at default.divFunc(Int64, Int64)(D:\git\cangjie_practice\书稿\code\Chapter12\dived_by_zero_demo.cj:2)
        at default.main()(D:\git\cangjie_practice\书稿\code\Chapter12\dived_by_zero_demo.cj:6)
```

执行除法操作的函数是 divFunc，这个函数看起来没什么问题，但是在实际执行中遇到了除数为 0 的情况，也就是参数 divisor 是 0，在这种情况下系统抛出了异常，于是程序不能正常执行下去。

针对函数 divFunc，在执行代码 dividend/divisor 时，可以预先判断 divisor 是否是 0，如果不是 0 就正常执行，如果是 0 就要做一些特殊处理，例如，可以返回一个特殊的值。也就是说，对于函数 divFunc，它的返回值其实有两种情况：一种是正常的情况，返回值是正确的结果；另一种是特殊的情况，返回值没有实际意义。仓颉语言的 enum 类型正好适合解决这种问题，内置的 Option 类型具有 Some 和 None 两个构造器，第 1 个构造器会携带一个参数，表示有值；第 2 个构造器不带参数，表示无值。函数 divFunc 正常返回值时，可以使用 Some 构造器包裹返回值，表示值是正常的；出现除数为 0 的异常时使用 None 构造器代表返回值，表示函数计算过程有问题，返回值不包括计算结果。Option 类型的定义如下：

```
public enum Option<T> {
    Some(T) | None
}
```

从定义可知，Option 类型是泛型的 enum，根据类型实参的不同可以表示各种数据类型的空值。使用 Option 改造后的示例代码如下：

```
//Chapter12/option_demo.cj

/*
    进行除法运算并返回结果
    dividend: 被除数
    divisor: 除数
    返回值: 运算结果,正常值使用 Some 包裹,运算异常时返回 None
*/
func divFunc(dividend: Int64, divisor: Int64): Option<Int64> {
    if (divisor == 0) {
        return Option<Int64>.None
    } else {
        return Option<Int64>.Some(dividend / divisor)
    }
}

func printValue(value: Option<Int64>) {
    match (value) {
        case None => println("Divided by Zero")         //返回值无值,表明计算出了异常
        case Some(v) => println("The result is ${v}")   //返回值有值,解构出值并打印
    }
}
main() {
    printValue(divFunc(100, 0))
    printValue(divFunc(100, 5))
}
```

编译后运行示例，输出如下：

```
Divided by Zero
The result is 20
```

在上述示例中，使用 Option<Int64>类型表示计算的结果，可以覆盖结果有值和无值的情况；从 Option<Int64>类型获取有效值时，使用 match 表达式，通过带变量的模式，可以把 Some 中包括的值提取到变量中，上例中 printValue 里 Option<Int64>包括的正常值就被提取到变量 v 中，在 Some 分支后的表达式中可以使用该变量。

Option 类型的写法可以简化为在类型名前加问号，例如对于 Option<Int64>类型可以简化为?Int64，两者是等价的，更多的示例代码如下：

```
let a: Option<Int64> = Some(100)
let b: ?Int64 = Some(100)

let c: Option<Char> = Some('C')
let d: ?Char = Some('C')
```

在某个位置明确需要 Option<T>类型时，可以直接将 T 类型传递过去，虽然两者的数据

类型不同,但是编译器都会用 Option<T>的 Some 构造器把 T 类型的值封装为 Option<T>类型的值,示例代码如下:

```
let a: Option<Int64> = 100
let b: ?Int64 = 100

let c: Option<Char> = 'C'
let d: ?Char = 'C'
```

### 12.5.4　区间类型

在 4.3.8 节介绍过区间类型,其实区间类型也是一个泛型,使用 Range<T> 表示。当 T 被实例化为不同的类型时(要求此类型必须支持关系操作符,并且可以和 Int64 类型的值做加法),会得到不同的区间类型,如最常用的 Range<Int64>用于表示整数区间。

# 第 13 章 包 管 理

在企业级的软件开发中,软件系统的代码量有可能达到几十万甚至上百万行的规模,在这种情况下,单个的代码文件无法满足软件开发和项目管理的要求,在实际开发中,一个软件项目包括众多的源代码文件,这些文件可以分组进行编译和管理。仓颉语言对源代码主要通过包和模块进行管理,在仓颉语言中包是编译的最小单元,一个包可以包括 1 个或者多个源代码文件;模块是第三方开发者发布的最小单元,一个模块可以包括 1 个或者多个包,下面详细讲解包和模块的用法。

## 13.1 包的声明

仓颉语言中包的声明如下:

```
package packageName
```

其中,package 为包声明的关键字,packageName 为包名,包名必须是一个合法的仓颉标识符。除了注释、空白字符和空行以外,包声明必须是源文件的首行,并且同一个包中的多个源文件的包声明必须保持一致。

在仓颉语言中,源代码根目录默认为 src,在源代码根目录下的包可以没有包声明,此时编译器会自动将包名称指定为 default。源代码根目录下其他文件夹内的源文件,其包名必须和所在文件夹相对于源代码根目录的相对路径保持一致(将路径分隔符替换为小数点),例如,对于图 13-1 所示的源代码目录结构,各个源文件的包声明如表 13-1 所示。

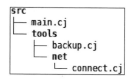

图 13-1 源代码目录结构

表 13-1 包声明对比

| 源 文 件 | 相对 src 的目录结构 | 包 声 明 |
| --- | --- | --- |
| main.cj | | 可以不显式声明,默认为 package default |
| backup.cj | tools | package tools |
| connect.cj | tools/net | package tools.net |

## 13.2　顶层声明的可见性

包的顶层声明在默认不使用可见性修饰符时仅包内可见,如果希望某个顶层声明被导出,在包外也可见,需要在此顶层声明前加上 public 修饰符。被 public 修饰的顶层声明不能使用包外不可见的类型。这一点可以这样理解,一个包外可见的顶层声明,在包外观察这个声明时,所涉及的所有类型必然也是外部可见的,否则就会出现未定义的类型。当然,在顶层声明或者定义内部,例如初始化表达式或者函数体内可以使用本包可见的类型,毕竟在包外观察不到这些类型。

## 13.3　包的导出和编译

一个包中的顶级声明或者定义被关键字 public 修饰后,可以被别的包导入使用,示例代码如下:

```
//Chapter13/export_type_demo.cj

package geometry

public interface Shape {
    func area(): Float64
}

public let PI: Float64 = 3.14

public class Rectangle <: Shape {
    public Rectangle(let width: Float64, let height: Float64) {}

    public func area() {
        return width * height
    }
}

public class Circle <: Shape {
    public Circle(let radius: Float64) {}

    public func area() {
        return PI * radius * radius
    }
}

//Chapter13/export_func_demo.cj

package geometry
```

```
public func compareShape(shape1: Shape, shape2: Shape) {
    if (shape1.area() > shape2.area()) {
        return 1
    } else if (shape1.area() < shape2.area()) {
        return -1
    } else {
        return 0
    }
}
```

该包的名称为 geometry,包括两个源文件 export_type_demo.cj 和 export_func_demo.cj,在源文件里定义了使用 public 修饰的接口、全局变量、类和函数。对包的编译可以使用 cjc 命令,格式如下:

```
cjc -p srcPath -o outPath --output-type=staticlib --module-name=moduleName
```

其中,cjc 是编译命令;-p 表示对包编译;srcPath 是源代码所在的路径;-o 指定输出路径;outPath 表示输出文件路径;--output-type=staticlib 表示输出静态链接库文件;--module-name=moduleName 表示指定模块名称为 moduleName。假如源文件在/data/code/demo/src/文件夹下,希望输出文件到/soft/cangjie/geo/,模块名称为 geo,则包编译命令如下:

```
cjc -p /data/code/demo/src/ -o /soft/cangjie/geo/ --output-type=staticlib --module-name=geo
```

编译后生成两个文件,分别是 geometry.cjo 和 libgeometry.a。

## 13.4 包的导入

### 13.4.1 import 语句导入

在使用其他包中的定义和声明时,首先需要导入其他包中的顶级声明或定义,语法如下:

```
from moduleName import packageName.itemName
```

其中,moduleName 为模块名,packageName 为包名,itemName 为声明的名字。导入标准库模块或当前模块中的内容时,可以省略 from moduleName;跨模块导入时,必须使用 from moduleName 指定模块。导入语句在源文件中的位置必须在包声明之后,并且在其他声明或定义之前。假如要导入 13.2 节中的接口 Shape,假设模块名称为 geo,导入操作的示例代码如下:

```
package Default

from geo import geometry.Shape
```

如果要导入一个包中所有被 public 修饰的顶级声明或定义,则可以使用 import

packageName.* 语法，如果要导入多个包，则需要用逗号分隔，示例代码如下：

```
from module_name import package1.*
from module_name import package1.*, package2.*
```

对于导入的声明或者定义，支持多种访问方式：
- moduleName.packageName.itemName
- packageName.itemName
- itemName

以 13.2 节中的类 Circle 为例，假如导入操作的示例代码如下：

```
package Default

from geo import geometry.*
```

那么，对 Circle 的实例化可以采取下面 3 种方式，示例代码如下：

```
let circle1 = geo.geometry.Circle(1.0)
let circle2 = geometry.Circle(1.0)
let circle3 = Circle(1.0)
```

通过导入调用 13.2 节中的定义的完整代码如下：

```
//Chapter13/import_demo.cj

package Default

from geo import geometry.*

main() {
    let result = geometry.compareShape(Rectangle(5.0, 6.0), Circle(3.0))
    match {
        case result > 0 => println("The first figure is larger than the second figure.")
        case result < 0 => println("The first figure is smaller than the second figure.")
        case _ => println("The two figures are the same size.")
    }
}
```

编译后运行示例，输出如下：

```
The first figure is larger than the second figure.
```

在编译时要注意，编译器会依次从当前路径、CANGJIE_PATH 指定的路径、CANGJIE_HOME 指定的路径下查询依赖的第三方包文件，要保证依赖的包在这些路径里。

在导入包时，要注意以下事项。

（1）只允许导入使用 public 修饰的顶级声明或定义，导入无 public 修饰的声明或定义将会在导入处报错。

（2）禁止包间的循环依赖导入，如果包之间存在循环依赖，则编译器会报错。

(3) 导入的声明或定义如果和当前包中的顶层声明或定义重名且不构成函数重载(关于函数重载可参考16.1节"函数重载"),则导入的声明和定义会被遮盖。

(4) 导入的声明或定义如果和当前包中的顶层声明或定义重名且构成函数重载,函数调用时将会根据函数重载的规则进行函数决议。

(5) 编译器会自动导入 core 包中所有 public 修饰的声明,例如 String、Range 等。

### 13.4.2 导入重命名

不同的包之间可能存在同名的顶级声明,为避免产生命名冲突,仓颉语言支持使用 import packageName.name as newName 的方式进行重命名;对于包名称冲突的情况,也支持使用 import pkg.* as newPkgName.* 的方式进行重命名,示例代码如下:

```
//模块名称为 module1
package p1

public func f1() {}

public class C1 {}


//模块名称为 module2
package p1

public func f1() {}

public class C2 {}
```

在上述示例中,有两个模块 module1 和 module2,这两个模块都有一个包叫作 p1,在新的模块 module3 中可以这样导入:

```
from module1 import p1.* as A.*
from module2 import p1.* as B.*
from module2 import p1.c2 as C
```

在包导入重命名时,还要注意作用域优先级的变化,假如被导入的包有一个 class 叫作 C,当前包也有一个 class 叫作 C,那么在使用 C 时可以把模块、包名都加上从而区分不同的 C;在不使用 import as 引入新的名字时,新的名字作用域优先级低于当前包顶层作用域,如果冲突,则引入的名字会被遮盖;如果使用 import as 引入新的名字,则新的名字作用域和当前包顶层作用域优先级一致,这时会提示名字冲突。

# 第 14 章 扩 展

## 14.1 扩展的定义

首先通过一个简短的示例了解什么是扩展,假如有这样一种需求,需要判断一个整型变量是否是偶数,通常的实现代码如下:

```
//Chapter14/even_num_demo.cj

//判断给定参数是否是偶数的函数
func isEvenNum(num: Int64) {
    if (num % 2 == 0) {
        return true
    } else {
        return false
    }
}

main() {
    var num = 5
    var result = isEvenNum(num)
    println(result)

    num = 6
    result = isEvenNum(num)
    println(result)
}
```

编译后运行该示例,输出如下:

```
false
true
```

这样,每次需要判断一个整数是否是偶数时,调用函数 isEvenNum 即可。那么,有没有其他方法呢?如果 Int64 类型有一个成员函数能判断实例是否是偶数就更方便了。仓颉语言提供了扩展功能,使用该功能在不破坏原有类型封装性的前提下,可以为当前 package 可

见的类型(除函数、元组、接口外)添加新功能,示例代码如下:

```
//Chapter14/extend_demo.cj

extend Int64 {
    func isEvenNum() {
        //this 指代扩展类型的实例
        if (this % 2 == 0) {
            return true
        } else {
            return false
        }
    }
}

main() {
    var num = 5
    var result = num.isEvenNum()
    println(result)

    num = 6
    result = num.isEvenNum()
    println(result)
}
```

编译后运行该示例,输出如下:

```
false
true
```

在上述示例中,给 Int64 类型添加了一个成员函数 isEvenNum,它用来判断类型实例是不是偶数,调用时很自然,和 Int64 本身的成员函数调用方式一样。

在定义扩展时,使用关键字 extend 声明,后面跟被扩展的类型,然后是一对大括号,大括号内是被添加的功能。可以添加的功能包括以下几种。

(1) 添加成员函数。

(2) 添加操作符重载函数。

(3) 添加成员属性。

(4) 实现接口。

因为扩展不破坏原有类型的封装性,所以扩展不支持以下功能。

(1) 扩展不能增加成员变量。

(2) 扩展的函数和属性必须拥有实现。

(3) 扩展的函数和属性不能使用 open、override、redef 修饰。

(4) 扩展不能访问原类型 private 的成员。

## 14.2 泛型扩展

泛型类型也支持扩展,扩展处的泛型变元与定义处的泛型变元的名称需保持一致,示例代码如下:

```
//Chapter14/generic_extend_demo.cj

class Item<T> {
    Item(let index: Int64, let content: T) {}
}

extend Item<T> {
    func getContent(): T {
        return content
    }
}

main() {
    let item = Item<Int64>(1, 1)
    println("index:${item.index} content:${item.getContent()}")
}
```

编译后运行该示例,输出如下:

```
index:1 content:1
```

泛型类型的扩展支持声明额外的泛型约束,在满足约束条件时可以使用扩展函数,示例代码如下:

```
//Chapter14/additional_constraints_demo.cj

class Item<T> {
    Item(let index: Int64, let content: T) {}
}

extend Item<T> where T <: ToString {
    func printItem() {
        println("index:${index} content:${content}")
    }
}

main() {
    let item = Item<Int64>(1, 1)
    item.printItem()
}
```

编译后运行该示例,输出如下:

```
index:1 content:1
```

在上述示例中，扩展泛型时添加了 ToString 的接口约束，所以可以在扩展函数里直接打印 content 的内容，在 main 函数中，泛型实参传入的是 Int64 类型，这种类型内置了对 ToString 接口的实现，所以可以调用 printItem 函数。

## 14.3　接口扩展

在添加一种类型的扩展功能时，可以直接添加，称为直接扩展，也可以在添加时实现接口，称为接口扩展，在同一个扩展内可以同时实现多个接口，多个接口之间使用 & 分隔，示例代码如下：

```
//Chapter14/interface_extend_demo.cj
interface Shape {
    func area(): Float64
}

class Rectangle {
    Rectangle(let width: Float64, let height: Float64) {}
}

//通过扩展同时实现了接口 Shape 和 ToString
extend Rectangle <: Shape & ToString {
    public func area() {
        return width * height
    }

    public func toString() {
        return "width: ${width},height: ${height}"
    }
}

main() {
    let shape: Shape = Rectangle(2.0, 3.0)
    println(shape.area())

    let str: ToString = Rectangle(3.0, 4.0)
    println(str)
}
```

编译后运行该示例，输出如下：

```
6.000000
width:3.000000,height:4.000000
```

上述示例中使用扩展为 Rectangle 类实现了两个接口 Shape 和 ToString。

在使用扩展给类型添加接口时，如果被扩展的类型已经包含了接口要求的函数，则在扩展时不需要重新实现，直接声明实现接口即可，不过，需要注意的是，当接口成员默认为

public 修饰时,也要保证被扩展类型的对应函数是由 public 修饰的,否则无法进行接口扩展,示例代码如下:

```
//Chapter14/declare_only_demo.cj

//接口声明了函数 area
interface Shape {
    func area(): Float64
}

//Rectangle 类包含 area 函数
class Rectangle {
    Rectangle(let width: Float64, let height: Float64) {}
    public func area() {
        return width * height
    }
}

//通过扩展声明 Rectangle 实现了接口 Shape
extend Rectangle <: Shape {}

main() {
    let shape: Shape = Rectangle(2.0, 3.0)
    println(shape.area())
}
```

编译后运行该示例,输出如下:

```
6.000000
```

# 第 15 章 基础集合类型

仓颉语言常用的基础集合类型包括 Array 和 ArrayList，这两种类型有特定的适用场景，本章将详细介绍它们的使用方法。

## 15.1 Array

### 15.1.1 Array 的定义

Array 是有序的单一元素的集合，被称为泛型数组，使用 Array＜T＞来表示，T 可以是任意类型，数组中的每个元素的类型都是 T。可以使用构造函数创建指定类型的 Array 实例，示例代码如下：

```
let arrayEmpty: Array<Int64> = Array<Int64>()
var arrayFloat64: Array<Float64> = Array<Float64>([1.0, 2.0, 3.14, 999.99])
let arrayString = Array<String>(["hello", "cangjie"])
```

在创建实例时，数组元素作为初始化参数被传递给构造函数，如果有多个元素，则元素之间使用逗号分隔，外层使用方括号[]包裹。允许定义没有元素的泛型数组，这种数组是空数组。

除了可以使用构造函数创建数组实例外，也可以直接使用字面量初始化数组，只需使用方括号将逗号分隔的值列表括起来，编译器会根据上下文自动推断数组字面量的类型，示例代码如下：

```
let arrayInt64 = [1,2,3,4,5]
let arrayString = ["hello", "cangjie"]
```

### 15.1.2 访问 Array

Array 是有序数组，对 Array 元素的访问常用的方式有以下两种。

**1. for-in 循环遍历**

当需要遍历 Array 所有的元素时，可以通过 for-in 循环获取每个元素对象，因为 Array

是有序的，所以每次遍历的顺序也是恒定的，示例代码如下：

```
//Chapter15/for_in_demo.cj

main() {
    let array = Array<Int64>([10, 9, 8, 7, 6])
    println("First traversal:")
    for (item in array) {
        println(item)
    }
    
    println("Second traversal:")
    for (item in array) {
        println(item)
    }
}
```

编译后运行该示例，输出如下：

```
First traversal:
10
9
8
7
6
Second traversal:
10
9
8
7
6
```

### 2. 下标语法访问

对于非空的 Array，可以使用下标访问指定位置的元素，Array 的元素下标序号是从 0 开始的，也就是说第 1 个元素的下标序号是 0，最后一个元素的下标序号是 Array 元素数量减去 1，获取 Array 元素个数的属性为 size，使用下标访问指定位置元素及遍历所有元素的示例代码如下：

```
//Chapter15/subscript_demo.cj

main() {
    let array = Array<String>(["Cangjie", "is", "the", "best", "programming", "language"])
    println(array[1])
    println(array[3])
    
    let size = array.size
    var index = 0
    while (index < size) {
        println(array[index])
```

```
            index++
        }
}
```

编译后运行该示例,输出如下:

```
is
best
Cangjie
is
the
best
programming
language
```

使用下标时需要注意下标不要越界,也就是出现下标小于 0、下标大于或等于元素数量的情况,如果下标越界,则会触发 IndexOutOfBoundsException 的运行时异常。

下标语法也支持范围访问,如果需要获取 Array 中的某一段元素,则可通过在下标中传入 Range 类型的值;当 Range 字面量在下标语法中使用时,可以省略 start 或 end;当省略 start 时,Range 会从 0 开始;当省略 end 时,Range 的 end 会延续到最后一位,示例代码如下:

```
//Chapter15/range_demo.cj

main() {
    let array = Array<String>(["Cangjie", "is", "the", "best", "programming", "language"])

    //包含下标为 0 和 1 的元素
    let arrayLeft = array[..2]

    //包含下标为 1、2、3、4 的元素
    let arrayMid = array[1..=4]

    //包含从下标 3 开始一直到最后的元素
    let arrayRight = array[3..]

    println("arrayLeft:")
    for (item in arrayLeft) {
        println(item)
    }

    println("arrayMid:")
    for (item in arrayMid) {
        println(item)
    }

    println("arrayRight:")
    for (item in arrayRight) {
```

```
        println(item)
    }
}
```

编译后运行该示例，输出如下：

```
arrayLeft:
Cangjie
is
arrayMid:
is
the
best
programming
arrayRight:
best
programming
language
```

## 15.1.3 修改 Array

Array 的长度不可改变，不允许添加或者删除元素，但是支持修改元素的值，通过下标语法定位到要修改的元素后，直接修改其值即可。Array 类型是引用类型，Array 类型变量的赋值或者作为函数参数等操作都不会复制副本，对元素的修改会影响所有指向该实例的引用，示例代码如下：

```
//Chapter15/modify_array_demo.cj

//把数组指定位置的元素值更新为新值
func updateArray(array: Array<Int64>, index: Int64, newValue: Int64) {
    array[index] = newValue
}

main() {
    let array = Array<Int64>([50, 60, 70, 80, 90, 100])
    let otherArray = array

    //直接通过下标给数组元素赋值
    array[1] = 66
    println("otherArray[1] = ${otherArray[1]}")

    //通过函数调用给数组指定元素赋值
    updateArray(otherArray, 2, 77)
    println("array[2] = ${array[2]}")
}
```

编译后运行该示例，输出如下：

```
otherArray[1] = 66
array[2] = 77
```

从上述示例可以看出,不管是把 Array 变量赋值给别的变量后修改元素,还是通过函数修改元素,都会影响所有指向该实例的变量。

### 15.1.4 Array 的高级用法

除了基本的访问和修改以外,Array 类型还提供了其他的用法,下面介绍常用的一些函数,更多的用法可参考仓颉库文档。

**1. 构造函数**

Array 提供了多个构造函数,常用的有以下两个。

- public init(size: Int64, item!: T)

使用初始值 item 初始化具有 size 个元素的数组,注意这里 item 是命名参数。

- public init(elements: Collection<T>)

使用 Collection 类型的元素 elements 创建数组。

使用的示例代码如下:

```
//Chapter15/init_array_demo.cj

main() {
    //创建包括 3 个 1024 数值元素的数组
    let array = Array<Int64>(3, item: 1024)

    //打印数组的每个元素
    println("array:")
    for (item in array) {
        println(item)
    }

    //使用已有数组创建一个新的数组
    let newArray = Array<Int64>(array)

    //打印新数组的每个元素,应该和原数组元素值一样
    println("newArray:")
    for (item in newArray) {
        println(item)
    }

    //将新数组的第 1 个元素值修改为 100
    newArray[0] = 100

    //输出原数组的第 1 个元素的值,检查是否还是 1024,从而证明原数组和新数组是两个不同的实例
    println("array[0] = ${array[0]}")
}
```

编译后运行该示例,输出如下:

```
array:
1024
```

```
1024
1024
newArray:
1024
1024
1024
array[0] = 1024
```

2．其他函数

- public func clone()：Array＜T＞

克隆数组，返回一个新的数组实例。

- public func isEmpty()：Bool

数组是否为空。

- public func reverse()：Unit

反转数组。

函数 reverse 使用的示例代码如下：

```
//Chapter15/reverse_array_demo.cj

main() {
    //创建数组
    let array = Array<Int64>([0, 1, 2, 3, 4, 5, 6])

    //生成新的克隆数组
    let cloneArray = array.clone()

    //反转克隆数组
    cloneArray.reverse()

    //打印原始数组的每个元素
    println("original array:")
    for (item in array) {
        println(item)
    }

    //打印克隆数组反转后的每个元素
    println("reverse array:")
    for (item in cloneArray) {
        println(item)
    }
}
```

编译后运行该示例，输出如下：

```
original array:
0
1
2
```

```
3
4
5
6
reverse array:
6
5
4
3
2
1
0
```

### 15.1.5 字节数组字面量

在进行字符处理时，可能需要对一系列字符的 ASCII 码进行运算，为了方便用 Array<UInt8>类型表示字符序列的 ASCII 码值，仓颉语言引入了字节数组字面量的概念，字节数组字面量由字符 b、一对用双引号引用的 ASCII 码字符串组成，示例代码如下：

```
var array = b"cangjie"
```

变量 array 表示一个长度为 7，类型为 Array<UInt8>的数组，其中的每个元素的值为对应字符的 ASCII 码的数值，即 [99,97,110,103,106,105,101]。字节数组字面量也支持转义字符，示例代码如下：

```
var array = b"\t\r\n"
```

这里，变量 array 表示一个长度为 3，类型为 Array<UInt8>的数组，元素值为[9,13,10]。

## 15.2 ArrayList

### 15.2.1 ArrayList 的定义

ArrayList 是可变的元素集合，使用 ArrayList<T>表示，T 可以是任何类型，列表中的每个元素的类型都是 T。可以使用构造函数创建指定类型的 ArrayList 实例，示例代码如下：

```
let emptyList = ArrayList<String>()
let intList = ArrayList<Int64>([0, 1, 2])
```

在创建实例时，集合元素作为初始化参数被传递给构造函数，如果有多个元素，则元素之间使用逗号分隔，外层使用方括号[]包裹。允许定义没有元素的泛型 ArrayList，这种列表是空 ArrayList。

### 15.2.2 访问 ArrayList

ArrayList 的访问方式和 Array 的访问方式类似，支持 for-in 循环遍历和下标语法访问,这里就不再给出示例了。

### 15.2.3 修改 ArrayList

与 Array 不同,ArrayList 不但支持修改某个位置的元素,而且可以增删元素。对特定位置元素的修改可以使用下标语法,示例代码如下:

```
let intList = ArrayList<Int64>([0, 1, 2])
intList[1] = 2
```

除了下标语法,其他主要的 ArrayList 修改函数如下。
- public func set(index: Int64, element: T): Unit

将 ArrayList 中 index 位置的元素替换为 element。
- public func append(element: T): Unit

将元素 element 添加到 ArrayList 末尾。
- public func appendAll(elements: Collection<T>): Unit

将集合 elements 中所有的元素添加到 ArrayList 末尾。
- public func insert(index: Int64, element: T): Unit

在 index 位置插入元素 element。
- public func insertAll(index: Int64, elements: Collection<T>): Unit

将集合 elements 中的所有元素从位置 index 开始插入。
- public func remove(index: Int64): T

删除 ArrayList 中 index 位置的元素。
- public func clear(): Unit

清空 ArrayList 中的所有元素。

下面通过一个具体的示例,演示 ArrayList 的常用函数的用法,示例代码如下:

```
//Chapter15/ArrayList_demo.cj

import collection.ArrayList

from std import collection.*
main() {
    let headList = ArrayList<Int64>([0, 1, 2])
    let tailList = ArrayList<Int64>([7, 8, 9])
    var targetList = ArrayList<Int64>()

    //依次给 targetList 添加元素 4、5、6
    targetList.append(4)
    targetList.append(5)
    targetList.append(6)

    //打印 targetList
```

```
    println(targetList)

    //在第 1 个位置插入 3
    targetList.insert(0, 3)

    //打印 targetList
    println(targetList)

    //将 headList 插入 targetList 开头
    targetList.insertAll(0, headList)

    //打印 targetList
    println(targetList)

    //将 tailList 添加到 targetList 末尾
    targetList.appendAll(tailList)

    //打印 targetList
    println(targetList)

    //删除第 1 个元素
    targetList.remove(0)

    //打印 targetList
    println(targetList)

    //反转集合元素
    targetList.reverse()

    //打印 targetList
    println(targetList)
}
```

编译后运行该示例,输出如下:

```
[4, 5, 6]
[3, 4, 5, 6]
[0, 1, 2, 3, 4, 5, 6]
[0, 1, 2, 3, 4, 5, 6, 7, 8, 9]
[1, 2, 3, 4, 5, 6, 7, 8, 9]
[9, 8, 7, 6, 5, 4, 3, 2, 1]
```

# 第 16 章 函数的进阶用法

在本书的第 5 章讲解了函数的基本用法,本章将继续讲解函数的进阶用法。

## 16.1 函数重载

### 16.1.1 函数重载的定义

7min

在一个作用域中,一个函数名可以对应多个函数定义,这种现象称为函数重载。仓颉语言中的函数重载还要考虑普通构造函数与主构造函数构成的重载,下面通过具体的示例演示多种类型的函数重载。

**1. 函数参数不同构成重载**

具有相同的函数名称,但是参数个数不同,或者参数个数相同,但是类型不同,这样的两个或者多个函数构成重载,示例代码如下:

```
func f(p: Int64): Unit {}

func f(p: String): String {
    return p
}

func f(p1: Int64, p2: Char): Unit {}

func f(p1: Int64, p2: Float64) {
    return p2
}
```

需要注意的是,函数返回值的类型是否相同不会影响重载,如果两个函数的名称和参数完全相同,但是返回值的类型不同,则这两个函数不构成重载,编译器会提示重载冲突。

**2. 参数不同的构造函数构成重载**

同一种类型内的构造函数,如果参数不同,则构成重载,示例代码如下:

```
class Rectangle {
    let width: Int64
    let height: Int64

    init(width: Int64, height: Int64) {
        this.width = width
        this.height = height
    }

    init(width: Int64) {
        this.width = width
        this.height = 1
    }

    init() {
        this.width = 1
        this.height = 1
    }
}
```

### 3. 参数不同的构造函数和主构造函数构成重载

仓颉语言中有主构造函数的存在,虽然主构造函数和 init 构造函数的名称不同,但是在认定重载时,两种构造函数按照同名的函数对待,在参数不同的情况下构成重载,示例代码如下:

```
class Rectangle {
    Rectangle(let width: Int64, let height: Int64) {}

    init(width: Int64) {
        this.width = width
        this.height = 1
    }
}
```

### 4. 定义在不同作用域的函数,在函数可见作用域中构成重载

函数定义在不同的作用域,并且名称相同但参数不同,如果在某一个作用域对这些函数都可见,则在这个作用域里函数也构成重载,示例代码如下:

```
func printVar(value: Int64) {
    println("outter: ${value}")
}

main() {
    func printVar(value: String) {
        println("inner: ${value}")
    }

    printVar(30)
}
```

### 5. 分别定义在父类和子类中的函数,在函数可见作用域中构成重载

父类中使用 public 和 protected 修饰的成员函数,或者缺省修饰符的成员函数,在子类中是可见的,这些函数可以与子类中同名但参数不同的函数构成重载,示例代码如下:

```
open class BaseClass {
    public open func baseFun(param: Int64) {}
}

class SubClass <: BaseClass {
    func baseFun(param1: Int64, param2: String) {}

    public override func baseFun(param: Int64) {}
}
```

在以上示例中需要注意的是子类中的函数 baseFun(param: Int64),它的名称和参数与父类中的函数完全一样,这时它不构成重载,而是覆盖了父类函数,所以需要使用 override 关键字修饰。

### 6. 不构成重载的场景

在同一个作用域内,如果两个函数同名但是不构成重载,编译器会报错,不构成重载的场景包括但不限于如下几种。

1) class、interface、struct 类型的静态成员函数和实例成员函数之间不能重载

如下例所示的代码编译时会报错(overloaded functions 'demoFun' cannot mix static and non-static):

```
class DemoClass {
    static func demoFun(param: Int64) {}

    func demoFun(param: String) {}
}
```

2) enum 类型的 constructor、静态成员函数和实例成员函数之间不能重载

如下例所示,编译时将报错(redefinition of declaration 'A'):

```
enum DemoEnum {
    A(Int64) | B(Int64) | C(String)

    func A(param: String): Unit {}
}
```

## 16.1.2 函数重载决议

在函数调用时,所有当前作用域下可见并且通过类型检查的函数构成一个候选集,候选集有多个函数时,会通过一系列的规则进行函数重载决议,最终决定调用哪个函数,规则如下。

### 1. 优先选择优先级高的作用域的函数

在多层嵌套情况下,越内层优先级越高,或者说越接近调用位置的作用域优先级越高,示例代码如下:

```
//Chapter16/overload_priority_demo.cj

open class BaseClass {}

class SubClass <: BaseClass {}

func printItem(item: SubClass) {
    println("global function!")
}

main() {
    func printItem(item: BaseClass) {
        println("inner function!")
    }

    printItem(SubClass())
}
```

编译后运行该示例,输出如下:

```
inner function!
```

在上述示例中,调用 printItem 函数时,全局函数 printItem(item: SubClass)和 main 函数内部的 printItem(item: BaseClass)都是适配的,但是内部函数距离调用处"更近",优先级更高,所以编译器选择的是 printItem(item: BaseClass)函数。

### 2. 最高优先级有多个函数,选择最匹配的

如果最高优先级的作用域仍然有多个函数可选,这时就找最匹配的那一个,示例代码如下:

```
//Chapter16/best_match_demo.cj

open class BaseClass {}

class SubClass <: BaseClass {}

main() {
    func printItem(item: BaseClass) {
        println("BaseClass function!")
    }

    func printItem(item: SubClass) {
        println("SubClass function!")
    }
}
```

```
        printItem(SubClass())
}
```

编译后运行该示例,输出如下:

```
SubClass function!
```

在上述示例中,调用 printItem 函数时,main 函数里的 printItem(item: BaseClass)和 printItem(item: SubClass)也都是适配的,但是,调用时的实参是 SubClass 类型,和 printItem(item: SubClass)的参数最匹配,所以实际决议时调用该函数。

**3. 子类和父类认为是同一作用域**

定义在父类和子类中的重载函数,计算作用域优先级时,按照同一作用域计算,示例代码如下:

```
//Chapter16/as_same_scope_demo.cj
open class BaseClass {
    func printItem(item: BaseClass) {
        println("BaseClass function!")
    }
}

class SubClass <: BaseClass {
    func printItem(item: SubClass) {
        println("SubClass function!")
    }
}

main() {
    let sub = SubClass()
    sub.printItem(BaseClass())
}
```

编译后运行该示例,输出如下:

```
BaseClass function!
```

在上述示例中,虽然使用子类的实例调用 printItem 函数,但是因为父类和子类是同一作用域,所以重载决议选择的是更匹配的父类中定义的函数。

## 16.2 函数遮盖

在仓颉语言中,不允许在同一个作用域内定义两个名称和参数完全相同的函数,那么在不同的作用域内是否可以呢?下面通过一个示例进行演示,代码如下:

```
//Chapter16/shadow_demo.cj
```

```
//全局函数
func printVar(value: Int64) {
    println("Global Print: ${value}")
}

main() {
    printVar(100)
    //嵌套函数
    func printVar(value: Int64) {
        println("Main Print: ${value}")
    }
    printVar(99)

    func demo() {
        //嵌套函数内的嵌套函数
        func printVar(value: Int64) {
            println("Inner Print: ${value}")
        }
        printVar(98)
    }
    demo()
}
```

编译后运行该示例,输出如下:

```
Global Print:100
Main Print:99
Inner Print:98
```

从上述示例可以看出,在不同作用域可以定义名称和参数完全相同的函数,并且高优先级作用域的函数会对低优先级的函数构成遮盖(在仓颉语言中称为 shadow)。具体分析上述示例,首先定义了一个全局的 printVar 函数,然后在 main 函数内定义了一个同名嵌套函数,最后在 main 函数内的 demo 函数里又定义了一个 printVar 函数。在 main 函数里第 1 次调用 printVar 时,在当前位置只能看到全局的 printVar 函数,所以输出 Global Print: 100;第 2 次调用的代码为 printVar(99),这时可见作用域内有两个 printVar 函数,因为 main 函数内的那个 printVar 函数的作用域比全局函数的作用域的优先级更高,所以调用该函数时输出为 Main Print:99;第 3 次调用是通过 demo 函数进行的,demo 函数内的作用域优先级高于前两个 printVar 函数所在的作用域,所以调用 demo 函数内的 printVar 函数时输出为 Inner Print:98。

## 16.3 操作符重载

### 16.3.1 操作符重载的必要性

假设有这样一种需求,在一个订单系统里,订单对象包括订单编号、订货数量、交货周期等成员变量,如果合并两个订单,就生成一个新的订单对象,这个新的订单对象的交货周期为原来两个订单中周期较短的那个,订单编号也是交货周期较短的那个订单的编号,订货数量为原来两个订单的订货数量的和,实现订单合并的示例代码如下:

```
//Chapter16/order_demo.cj

//订单类
class Order <: ToString {
    /*
        构造函数
        orderNo: 订单号
        count: 订货数量
        supplyCycle: 交货周期
    */
    Order(var orderNo: String, var count: Int64, var supplyCycle: Int64) {}

    public func toString() {
        return "OrderNo: ${orderNo} Count: ${count} SupplyCycle: ${supplyCycle}"
    }
}

//合并两个订单
func mergeOrder(order1: Order, order2: Order): Order {
    let order = Order("", 0, 0)

    //判断交货周期较短的订单,新订单的订单号和交货周期设置为该订单的订单号和交货周期
    if (order1.supplyCycle <= order2.supplyCycle) {
        order.supplyCycle = order1.supplyCycle
        order.orderNo = order1.orderNo
    } else {
        order.supplyCycle = order2.supplyCycle
        order.orderNo = order2.orderNo
    }

    //合并订货数量
    order.count = order1.count + order2.count
    return order
}

main() {
    let order1 = Order("20221001001", 100, 30)
    let order2 = Order("20221001101", 356, 20)
    let newOrder = mergeOrder(order1, order2)
    println(newOrder)
}
```

编译后运行该示例,输出如下:

OrderNo:20221001101 Count:456 SupplyCycle:20

在上述示例中,通过函数 mergeOrder 实现了两个订单的合并,这从功能上没有任何问题,但是,如果能使用加号(+)直接对两个订单对象进行操作,则应该更自然、更方便,这在仓颉语言里称为操作符重载,针对上述的示例,可以对代码进行改造,改造后的代码如下:

```
//Chapter16/operators_overloading_demo.cj

//订单类
class Order <: ToString {
    /*
        构造函数
        orderNo: 订单号
        count: 订货数量
        supplyCycle: 交货周期
    */
    Order(var orderNo: String, var count: Int64, var supplyCycle: Int64) {}

    public func toString() {
        return "OrderNo: ${orderNo} Count: ${count} SupplyCycle: ${supplyCycle}"
    }

    //对 Order 类重载 + 操作符
    operator func + (right: Order): Order {
        let order = Order("", 0, 0)

        if (supplyCycle <= right.supplyCycle) {
            order.supplyCycle = supplyCycle
            order.orderNo = orderNo
        } else {
            order.supplyCycle = right.supplyCycle
            order.orderNo = right.orderNo
        }

        order.count = count + right.count
        return order
    }
}

main() {
    let order1 = Order("20221001001", 100, 30)
    let order2 = Order("20221001101", 356, 20)
    let newOrder = order1 + order2
    println(newOrder)
}
```

编译后运行该示例,输出如下:

```
OrderNo:20221001101 Count:456 SupplyCycle:20
```

使用操作符重载,可以简化代码,提高代码的可读性。

## 16.3.2 操作符重载的定义

操作符重载函数和普通函数的定义类似,函数名称为要重载的操作符,在 func 关键字前需要添加 operator 修饰符。定义操作符函数时还要注意以下几点。

(1) 操作符函数的参数个数和操作符要求的操作数个数必须相同。
(2) 操作符函数只能定义在 class、interface、struct、enum 和 extend 中。
(3) 操作符函数具有实例成员函数的语义,禁止使用 static 修饰符。
(4) 操作符函数不能为泛型函数。
(5) 被重载后的操作符不改变它们固有的优先级和结合性。

操作符函数的定义有两种方式,第 1 种针对可以直接包含函数定义的类型(包括 struct、enum、class 和 interface),直接在其内部定义,详细定义示例可参考 16.3.1 节展示的 //Chapter16/ operators_overloading_demo. cj 代码文件。第 2 种针对无法直接包含函数定义的类型(指除 struct、class、enum 和 interface 之外的其他类型)或无法改变其实现的类型,例如第三方定义的 struct、class、enum 和 interface,可以使用扩展的方式为其添加操作符函数,示例代码如下:

```
//Chapter16/plus_operator_demo.cj

from std import collection.*

//扩展 ArrayList
extend ArrayList<T> {
    //重载 + 操作符,使 ArrayList 类型支持 + 运算
    operator func +(right: ArrayList<T>): ArrayList<T> {
        //把原 ArrayList 克隆一份,得到一个新的 ArrayList
        let newList = this.clone()

        //把右侧的 ArrayList 追加到新的 ArrayList 的后面
        newList.appendAll(right)

        //返回追加后的新 ArrayList
        return newList
    }
}

main() {
    //定义两个 ArrayList
    let List1 = ArrayList<Int64>([1, 2, 3])
    let List2 = ArrayList<Int64>([4, 5, 6])

    //使用重载的 + 操作符进行运算
    let List3 = List1 + List2
```

```
    //打印列表
    println(List1)
    println(List2)
    println(List3)
}
```

编译后运行该示例,输出如下:

```
[1, 2, 3]
[4, 5, 6]
[1, 2, 3, 4, 5, 6]
```

### 16.3.3 索引操作符

索引操作符([])是一种特殊的操作符,常用于类似 Array 这种集合对象,其他操作符重载时,一般提供一个重载函数即可,而索引操作符具有取值和赋值两种形式,可以提供两个重载函数,分别如下:

```
operator func [](index: D): T {
    ...
}

operator func [](index: D, value: T): Unit {}
```

其中,index 表示下标序号,value 表示要赋予的值,索引操作符重载函数的示例代码如下:

```
//Chapter16/index_operator_demo.cj

//定义火车类
class Train {
    //定义数组类型的车厢变量,每节车厢放一个数字
    let carriage: Array<Int64> = Array<Int64>([1, 2, 3, 4, 5, 6, 7, 8, 9])

    //获取指定位置车厢中的数字
    operator func [](index: Int64): Int64 {
        return carriage[index]
    }

    //设置指定位置车厢中的数字
    operator func [](index: Int64, value: Int64): Unit {
        carriage[index] = value
    }
}

main() {
    //定义火车变量
    let train = Train()
```

```
    //将索引为 3 处的车厢数字设置为 1024
    train[3] = 1024

    //打印索引为 3 处的车厢数字
    println(train[3])
}
```

编译后运行该示例,输出如下:

```
1024
```

### 16.3.4 可以被重载的操作符

仓颉语言不支持自定义操作符,所有可以被重载的操作符如表 16-1 所示,按照优先级从高到低排列。

表 16-1 可以被重载的操作符

| 操作符 | 说明 | 操作符 | 说明 |
| --- | --- | --- | --- |
| () | 函数调用 | >> | 按位右移 |
| [] | 索引 | < | 小于 |
| ! | 逻辑非 | <= | 小于或等于 |
| - | 一元负号 | > | 大于 |
| ** | 幂运算 | >= | 大于或等于 |
| * | 乘法 | == | 判等 |
| / | 除法 | != | 判不等 |
| % | 取模 | & | 按位与 |
| + | 加法 | ^ | 按位异或 |
| - | 减法 | \| | 按位或 |
| << | 按位左移 | | |

## 16.4 函数是第一类对象

### 16.4.1 什么是第一类对象

13min

在编程语言中,如果一个实体可以支持其他实体通常可用的所有操作,包括参数传递、从函数返回及赋值给变量等,则这个实体可以认为是第一类对象。在大部分编程语言中,基本数据类型及结构体、枚举、类、接口等实体都被认为是第一类对象,但是对于函数,各种语言的设计是不同的,传统的如 C、Java 等语言,没有把函数作为第一类对象;一些比较现代的语言,如 Scala、Go 等把函数作为第一类对象。

在仓颉语言中,函数可以作为函数的参数或返回值,也可以赋值给变量,是第一类对象,它本身也有类型,被称为函数类型。

注意:"第一类对象"(First-class object)原来也被称为"一等公民"(First-class citizen),

由克里斯托弗·斯特雷奇在20世纪60年代发明,英文中也称为First-class entity 或 First-class value。

本节部分内容参考引用了维基百科,网址为 https://zh.wikipedia.org/,依据"CC BY-SA 3.0"许可证进行授权。要查看该许可证,可访问 https://creativecommons.org/licenses/by-sa/3.0/。

### 16.4.2 函数类型的定义

函数类型由函数的参数类型、参数数量、参数顺序及返回类型决定,格式如下:

```
(ParamType1,ParamType2,…) -> ReturnType
```

其中,小括号内的 ParamType1、ParamType2 为函数的参数类型,参数类型之间使用逗号分隔,函数类型可以拥有 0 个、1 个或者多个参数类型;ReturnType 为返回值类型,小括号和 ReturnType 之间使用一>连接。各种函数类型的示例如下。

**1. 无参数函数类型**

```
func noParamFunc(): Unit {
    println("Hello cangjie!")
}
```

该实例没有参数,返回类型为 Unit,函数类型为

```
() -> Unit
```

**2. 单参数函数类型**

```
func oneParamFunc(value: String): Unit {
    println(value)
}
```

该实例有一个参数,返回类型为 Unit,函数类型为

```
(String) -> Unit
```

**3. 多参数函数类型**

```
func buildTupleFunc(name: String, age: Int64): (String,Int64) {
    (name, age)
}
```

该实例有两个参数,返回类型为元组,函数类型为

```
(String,Int64) -> (String,Int64)
```

**4. 返回自定义类型函数类型**

```
class Circle {
    Circle(let radis: Int64) {}
}
```

```
func getCircleArray(radis1: Int64, radis2: Int64): Array<Circle> {
    return Array<Circle>([Circle(radis1), Circle(radis2)])
}
```

该实例有两个参数,返回类型为自定义类 Circle 的数组,函数类型为

```
(Int64, Int64) -> Array<Circle>
```

### 16.4.3 函数作为参数

函数是第一类对象,函数类型可以像其他类型一样,作为参数传递给函数,示例代码如下:

```
//Chapter16/function_parameter_demo.cj

//使用汉语输出欢迎语
func printWithChinese(name: String) {
    println("${name},您好")
}

//使用英语输出欢迎语
func printWithEnglish(name: String) {
    println("Hello, ${name}")
}

//使用给定的函数参数 printFunc 输出 name
func welcome(name: String, printFunc: (String) -> Unit) {
    printFunc(name)
}

main() {
    var name = "小明"
    welcome(name, printWithChinese)

    name = "Tom"
    welcome(name, printWithEnglish)
}
```

编译后运行该示例,输出如下:

```
小明,您好
Hello,Tom
```

在这个示例中,定义了两个(String)->Unit 类型的函数,这两个函数在 main 函数里作为参数传递给了 welcome 函数。

### 16.4.4 函数作为变量

定义一个函数类型的变量后,可以把该函数类型的函数赋值给该变量,并且可以直接通过函数变量执行函数,示例代码如下:

```
//Chapter16/function_variable_demo.cj

//使用汉语输出欢迎语
func printWithChinese(name: String) {
    println(" ${name},您好")
}

//使用英语输出欢迎语
func printWithEnglish(name: String) {
    println("Hello, ${name}")
}

main() {
    //定义函数变量
    var printFun: (String) -> Unit
    var name = "小明"

    //把函数名称赋值给函数变量
    printFun = printWithChinese

    //使用函数变量调用函数
    printFun(name)

    name = "Tom"
    printFun = printWithEnglish
    printFun(name)
}
```

编译后运行该示例,输出如下:

```
小明,您好
Hello,Tom
```

## 16.4.5　函数作为返回值

函数的返回类型也可以定义为函数类型,示例代码如下:

```
//Chapter16/function_return_type_demo.cj

//使用汉语输出欢迎语
func printWithChinese(name: String) {
    println(" ${name},您好")
}

//使用英语输出欢迎语
func printWithEnglish(name: String) {
    println("Hello, ${name}")
}
```

```
//根据编码获取对应的输出函数
func getPrintFunc(code: String): (String) -> Unit {
    if (code == "CN") {
        return printWithChinese
    } else {
        return printWithEnglish
    }
}

main() {
    //定义函数变量
    var printFun: (String) -> Unit
    var name = "小明"

    //根据编码获取输出函数
    printFun = getPrintFunc("CN")
    printFun(name)

    name = "Tom"

    //根据编码获取输出函数并直接调用该函数
    getPrintFunc("EN")(name)
}
```

编译后运行该示例,输出如下:

```
小明,您好
Hello,Tom
```

# 第 17 章 类 型 关 系

## 17.1 多态

多态是面向对象开发中的重要概念,根据维基百科的解释,多态指"为不同数据类型的实体提供统一的接口,或使用一个单一的符号来表示多个不同的类型"。在前述章节中,其实已经应用了多态的概念,例如,在定义函数时,可以把形参的类型定义为一个父类,在调用时,给实参传递子类的实例即可。在定义接口的实现时,可以在不同的类型里实现同一个接口的成员。多态的示例代码如下:

```
//Chapter17/polymorphic_demo.cj

from std import math.*

//定义父类
open class Parent {
    public open func printInfo() {
        println("This is parent!")
    }
}

//定义子类
class Child <: Parent {
    public override func printInfo() {
        println("This is child!")
    }
}

//定义函数,形参为父类
func callPrint(obj: Parent) {
    obj.printInfo()
}

//定义接口
interface Figure {
```

```
    //获取周长的成员函数
    func perimeter(): Float64
}

//实现了接口 Figure 的矩形
class Rectangle <: Figure {
    Rectangle(let width: Float64, let height: Float64) {}

    //计算周长的函数
    public func perimeter(): Float64 {
        return 2.0 * (width + height)
    }
}

//实现了接口 Figure 的圆形
class Circle <: Figure {
    Circle(let radius: Float64) {}

    //计算周长的函数
    public func perimeter(): Float64 {
        return 2.0 * Float64.PI * radius
    }
}

//定义函数,形参为接口
func printPerimeter(figure: Figure) {
    println("The perimeter of the figure is ${figure.perimeter()}")
}

main() {
    callPrint(Parent())
    callPrint(Child())
    printPerimeter(Rectangle(1.0, 2.0))
    printPerimeter(Circle(2.0))
}
```

编译后运行该示例,输出如下:

```
This is parent!
This is child!
The perimeter of the figure is 6.000000
The perimeter of the figure is 12.566371
```

在本例中,多态产生的本质原因是子类型关系,类 Child 是 Parent 的子类型,类 Rectangle 和 Circle 是接口 Figure 的子类型;以函数 printPerimeter 为例,在程序运行时才可以最终确定实参的实际类型,也就是在运行时才能知道最终调用的 perimeter 函数究竟是 Rectangle 的成员还是 Circle 的成员。了解仓颉语言中哪些类型关系是子类型,有助于进一步理解多态的用法。

## 17.2 子类型关系

### 17.2.1 继承带来的子类型关系

在 class 类型中,子类对父类的继承形成了子类型关系,子类即为父类的子类型。以如下的代码为例,class C 继承了 class P,那么 C 就是 P 的子类型,代码如下:

```
open class P {}
class C <: P {}
```

### 17.2.2 实现接口带来的子类型关系

当某种类型 T 实现了接口 I1 后,类型 T 就成为接口 I1 的子类型;当某种类型 E 通过扩展实现了接口 I2 后,也认为类型 E 是接口 I2 的子类型,示例代码如下:

```
interface I1 {}
class T <: I1 {}

interface I2 {}
class E {}
extend E <: I2 {}
```

### 17.2.3 元组类型的子类型关系

如果一个元组 t1 的每种类型都是另一个元组 t2 对应位置的类型的子类型,则元组 t1 的类型也是元组 t2 的类型的子类型,示例代码如下:

```
open class P1 {}
open class P2 {}

class C1 <: P1 {}
class C2 <: P2 {}

let tupleItem1: (P1, P2) = (C1(), C2())
let tupleItem2: (P2, P1) = (C2(), C1())
```

### 17.2.4 函数类型的子类型关系

在仓颉语言中,函数作为第一类对象,也有自己的类型,叫作函数类型,在第 16 章已经介绍了函数类型及其用法。与本节已介绍的其他子类型关系类似,函数类型也有自己的子类型,假设有 4 种类型 P1、P2、C1、C2,其中 C1 <: P1,C2 <: P2,那么对于函数类型(C1)-> P2 和(P1)-> C2 有以下关系:(C1)-> P2 <: (P1)-> C2,也就是说函数类型(C1)-> P2 是(P1)-> C2 的子类型,在使用函数类型(P1)-> C2 的地方可以替换为(C1)-> P2 类型,示例代码如下：

```
//Chapter17/func_type_demo.cj

open class Meat {
    Meat(let name: String) {}
}

//Beef 是 Meat 的子类型
class Beef <: Meat {
    Beef(name: String) {
        super(name)
    }
}

open class Food {}

//RoastedMeat 是 Food 的子类型
class RoastedMeat <: Food {}

//形参是 Meat 类型，返回值是 Food 的子类型 RoastedMeat
func cook(ingredients: Meat): RoastedMeat {
    println("Cooking food,The ${ingredients.name} is delicious")
    return RoastedMeat()
}

//形参是 Meat 类型的子类型 Beef，返回值是 Food 类型
func barbecue(ingredients: Beef): Food {
    println("Grilling food,The ${ingredients.name} is delicious")
    return Food()
}

main(): Unit {
    //定义和 barbecue 函数类型一致的函数变量
    var cookFunc: (Beef) -> Food = barbecue
    cookFunc(Beef("beef"))

    //把 barbecue 函数类型的子类型 cook 赋给变量
    cookFunc = cook
    cookFunc(Beef("beef"))
}
```

编译后运行该示例，输出如下：

```
Grilling food,The beef is delicious
Cooking food,The beef is delicious
```

提示：函数类型的子类型关系对于初学者理解起来有一定的难度，在通常的软件开发

中应用也较少，本节内容只需做初步了解。

### 17.2.5 预设子类型关系

（1）一种类型 T 永远是自身的子类型，即 T <: T。
（2）Nothing 类型永远是其他任意类型 T 的子类型，即 Nothing <: T。
（3）任意类型 T 都是 Any 类型的子类型，即 T <: Any。
（4）任意 class 定义的类型都是 Object 的子类型，即如果有 classC{}，则 C <: Object。

### 17.2.6 传递性带来的子类型关系

假如有类型 T1 <: T2、T2 <: T3，那么有 T1 <: T3，也就是说，子类型具有传递性。

### 17.2.7 泛型类型的子类型关系

泛型类型也有子类型关系，如果定义了一个泛型接口，则实现了该泛型接口的类型就是泛型接口类型的子类型，示例代码如下：

```
//Chapter17/generic_subtype_demo.cj

//定义泛型接口
interface Printable<T> {
    func printInfo(item: T): Unit
}

//定义实现了泛型接口的泛型类
class Print<T> <: Printable<T> where T <: ToString {
    public func printInfo(item: T): Unit {
        println(item)
    }
}

//定义泛型函数,泛型接口为形参
func printItem<T>(item: Printable<T>, info: T) {
    item.printInfo(info)
}

main() {
    //使用 Int64 类型的泛型接口定义变量
    let int64Print: Printable<Int64> = Print<Int64>()

    //把泛型接口变量传递给函数
    printItem(int64Print, 100)

    //使用布尔类型的泛型接口定义变量
    let boolPrint: Printable<Bool> = Print<Bool>()

    //把泛型接口变量传递给函数
```

```
        printItem(boolPrint, true)
}
```

编译后运行该示例,输出如下:

```
100
true
```

在上述示例中,使用泛型接口作为形参的函数都可以使用实现了该接口的泛型类型作为实参传入。

## 17.3 类型转换

### 17.3.1 is 操作符

在 4.4.3 节"类型判断"中讲解了 is 操作符的基本用法,可以用来判断表达式是不是某个基本类型;除了基本类型以外,is 操作符也可以用来判断自定义类型是不是指定的类型或其子类型,示例代码如下:

```
//Chapter17/operator_is_demo.cj

//定义接口
interface Figure {}

//实现了接口的类
open class Circle <: Figure {
    Circle(let radius: Float64) {}
}

//继承自 Circle 的类
class ColorCircle <: Circle {
    ColorCircle(radius: Float64, let color: String) {
        super(radius)
    }
}

main() {
    //定义 ColorCircle 类型的变量 redCircle
    let redCircle = ColorCircle(1.0, "red")

    //变量 redCircle 是不是 Figure 类型
    println("redCircle is Figure: ${redCircle is Figure}")

    //变量 redCircle 是不是 Circle 类型
    println("redCircle is Circle: ${redCircle is Circle}")

    //变量 redCircle 是不是 Int64 类型
    println("redCircle is Int64: ${redCircle is Int64}")
}
```

编译后运行该示例，输出如下：

```
redCircle is Figure:true
redCircle is Circle:true
redCircle is Int64:false
```

### 17.3.2　as 操作符

当需要把一个表达式转换为指定的类型时，可以使用 as 操作符，格式如下：

```
e as T
```

其中，e 可以是任意表达式，T 可以是任何类型，当 e 的运行时类型是 T 的子类型时，e as T 的返回值为 Option<T>.Some(e)，否则返回 Option<T>.None，示例代码如下：

```
//Chapter17/operator_as_demo.cj

//定义接口
interface Figure {}

//实现了接口的类
open class Circle <: Figure {
    Circle(let radius: Float64) {}
}

//继承自 Circle 的类
class ColorCircle <: Circle {
    ColorCircle(radius: Float64, let color: String) {
        super(radius)
    }
}

main() {
    let colorRed = ColorCircle(1.0, "red")

    //把 colorRed 转换为 Circle 类型
    let circle = (colorRed as Circle).getOrThrow()

    //打印转换后的 radius 变量
    println(circle.radius)

    //把浮点型变量 circle.radius 转换为整型
    let optInt = circle.radius as Int64

    //判断转换是否成功
    match (optInt) {
        case Some(value) => println(value)
        case None => println("optInt is not of type Int64!")
    }
}
```

```
    //把circle转换为IFigure接口类型
    let optFigure = circle as Figure

    //判断转换是否成功
    match (optFigure) {
        case Some(value) => println("optFigure is a figure!")
        case None => println("optFigure is not a figure!")
    }
}
```

编译后运行该示例,输出如下:

```
1.000000
optInt is not of type Int64!
optFigure is a figure!
```

## 17.4 类型别名

仓颉语言支持定义类型别名,当某种类型的名字比较复杂或者在特定场景中不够直观时,可以选择使用类型别名的方式为此类型设置一个别名,定义格式如下:

```
type NewType = OriType
```

其中,type 是关键字,OriType 是原类型,NewType 是类型别名,示例代码如下:

```
type Short = Int16
```

在上述示例中,使用 Short 作为原类型 Int16 的别名。

在类型别名的定义时,要注意以下几点。

(1) 只能在源文件顶层定义类型别名,并且原类型必须在别名定义处可见。

(2) 在一个(或多个)类型别名定义中禁止出现(直接或间接的)循环引用。

(3) 类型别名不是新的类型,别名和原类型被视作同一种类型,可以像原类型一样使用。

类型别名的示例代码如下:

```
//Chapter17/type_alias_demo.cj

type Short = Int16
type Long = Int32

main() {
    let numShort: Short = 100
    let numLong: Long = 100

    let numInt16: Int16 = 10
    let numInt32: Int32 = 10
```

```
    println(numShort + numInt16)
    println(numLong + numInt32)
}
```

编译后运行该示例,输出如下:

```
110
110
```

# 第 18 章 异常

在程序的实际运行过程中，可能会有一些因素导致程序不能正常工作，例如磁盘空间不足、设备损坏、网络突然中断、除以 0 引起的溢出等，这些不被期望的事件统称为异常（Exception）。异常的出现具有偶然性，但是造成的后果有可能比较严重，一旦异常发生，程序可能终止，或者运行的结果不可预知。

仓颉语言提供了内置的异常处理机制用于处理程序运行时可能出现的各种异常情况，在异常发生时，可以捕获异常，将程序的控制权从正常功能的执行处转移至处理异常的部分，从而保证系统的正确性、稳定性和稳健性。

## 18.1 异常的定义

在仓颉语言中，异常有两个基类，分别是 Exception 和 Error。

（1）Error 类描述仓颉语言运行时，系统内部错误和资源耗尽错误，应用程序不应该抛出这种类型的错误，如果出现内部错误，则只能通知给用户，尽量安全终止程序。

（2）Exception 类描述的是程序运行时的逻辑错误或者 IO 错误导致的异常，例如数组越界或者试图打开一个不存在的文件等，这类异常需要在程序中捕获处理。

Error 是所有错误类的基类，该类不可被继承，不可初始化，可以被捕获到；Exception 是所有异常类的基类，用户可以通过继承 Exception 类或其子类来自定义异常。

Error 类的主要成员说明如表 18-1 所示。

表 18-1 Error 的成员

| 成 员 | 说 明 |
| --- | --- |
| public open prop message: String | 错误信息 |
| public open func toString(): String | 返回错误信息 |
| public func printStackTrace(): Unit | 打印堆栈信息 |
| public func getStackTrace(): Array<StackTraceElement> | 得到堆栈信息数组 |
| protected open func getClassName(): String | 获得类名 |

Exception 类的主要成员说明如表 18-2 所示。

表 18-2　Exception 的成员

| 成　　员 | 说　　明 |
| --- | --- |
| public init() | 构造函数 |
| public init(message: String) | 构造函数，message 为异常信息 |
| public open prop message: String | 异常信息 |
| public open func toString(): String | 返回异常信息 |
| public func printStackTrace(): Unit | 打印堆栈信息 |
| public func getStackTrace(): Array<StackTraceElement> | 得到堆栈信息数组 |
| protected open func getClassName(): String | 获得类名 |

## 18.2　异常处理

在程序中对异常的处理是通过 try 表达式完成的，try 表达式一般包括 3 部分，分别是 try 块、catch 块和 finally 块。下面通过一个具体的示例演示异常的处理，代码如下：

```
//Chapter18/exception_demo.cj

from std import convert.*

main() {
    //定义被除数变量并初始化为 0
    let divisor = Int64.parse("0")
    try {
        //进行除以 0 的运算，用来触发异常
        let quotient = 100 / divisor
        println(quotient)
    } catch (e: ArithmeticException) {
        //捕获算术异常
        println("An arithmetic exception occurred: ${e.message}")
    } catch (e: Exception) {
        //捕获异常
        println("An exception occurred: ${e.message}")
    } finally {
        println("This part is always executed")
    }
}
```

编译后运行该示例，输出如下：

```
An arithmetic exception occurred:Divided by zero !
This part is always executed
```

在上述示例中，关键字 try 及其后面大括号中的代码一起构成了 try 块；try 块后面是多个 catch 块，catch 块以关键字 catch 开头，后跟使用小括号括起来的匹配项，匹配项后面大括号内是 catch 块的代码;最后是一个 finally 块,finally 块以关键字 finally 开头,后面大括号内是 finally 块的代码。一个 try 表达式包括一个 try 块,0 个或者多个 catch 块,0 个或者一个 finally 块,但是 catch 块和 finally 块至少有一个。发生异常时,catch 块会通过模式匹配的方式捕获异常,如果某个异常的类型是 catch 块匹配项的子类,则该异常就会被这个 catch 块捕获,它后面的所有 catch 块都会被忽略。无论是否发生异常,finally 块中的代码都会被执行,如果发生异常但是没有被处理,则在执行完 finally 块后,异常会继续被抛出。

## 18.3 自定义异常

除了仓颉语言内置的异常类以外,开发者也可以根据需要自定义异常类,并在需要时抛出异常。下面通过一个具体的示例演示自定义异常处理,假设有一个进行整数除法运算的函数 integerDivision,该函数在除数为 0 时会创建并抛出一个除数为 0 的异常 DivisorIsZeroException,示例代码如下:

```
//Chapter18/user_defined_exception_demo.cj

//自定义异常类
class DivisorIsZeroException <: Exception {
    init() {
        super("DivisorIsZeroException Divisor is zero!")
    }
}

//除运算函数,在除数为 0 时抛出自定义异常 DivisorIsZeroException
func integerDivision(dividend: Int64, divisor: Int64): Int64 {
    if (divisor == 0) {
        throw DivisorIsZeroException()
    }

    return dividend / divisor
}

main() {
    try {
        let quotient = integerDivision(100, 0)
        println(quotient)
    } catch (e: DivisorIsZeroException) {
        //首先匹配并捕获自定义异常 DivisorIsZeroException
        println("An DivisorIsZero exception occurred: ${e.message}")
    } catch (e: Exception) {
        //其他异常不匹配时再匹配基础的 Exception 类
        println("An exception occurred: ${e.message}")
    }
}
```

编译后运行该示例,输出如下:

```
An DivisorIsZero exception occurred:DivisorIsZeroException Divisor is zero!
```

## 18.4 Option 值的解构

Option 类型是一个常用的类型,在 12.5.3 节"泛型 enum"里介绍了解构 Option 类型的一种方式,即通过模式匹配获取 Option 类型的值,本节将介绍另外两种解构方式。

### 18.4.1 getOrThrow()函数

对于 Option<T>类型的表达式 e,可以通过调用 getOrThrow()成员函数实现解构,当 e 的值等于 Some(v)时,getOrThrow()返回 v 的值,否则抛出异常,示例代码如下:

```
//Chapter18/option_get_or_throw_demo.cj

main() {
    let a: Option<Int64> = 1
    let b: ?Int64 = None

    try {
        let value = a.getOrThrow()
        println("The value of a is ${value}!")
    } catch (e: Exception) {
        println("The value of a is none!")
    }

    try {
        let value = b.getOrThrow()
        println("The value of b is ${value}!")
    } catch (e: Exception) {
        println("The value of b is none!")
    }
}
```

编译后运行该示例,输出如下:

```
The value of a is 1!
The value of b is none!
```

### 18.4.2 ?? 操作符

对于 Option<T>类型的表达式 e,使用??操作符时,如果 e 的值为 Some(v),则返回 v,如果 e 的值为 None,则返回??操作符右侧同样为 T 类型的操作数,示例代码如下:

```
//Chapter18/coalescing_demo.cj
```

```
main() {
    let a: Option<Int64> = 1
    let b: ?Int64 = None

    var value = a??0
    println(value)

    value = b??0
    println(value)
}
```

编译后运行该示例,输出如下:

```
1
0
```

# 第 19 章 基础类库

仓颉语言提供了丰富的内置类库，了解并恰当使用这些类库可以简化开发者的开发工作。为了方便后续章节的讲解和演示，本章将介绍部分类库的常用功能，更多的类库可以参考仓颉相关文档。

## 19.1 格式化库

格式化库用来支持格式化的输出，位于 std 模块的 format 包下，导入方式如下：

```
from std import format.*
```

目前提供的接口为 Formatter，继承该接口的类型主要有整型、浮点型及 char 类型，接口定义如下：

```
public interface Formatter {
    func format(fmt: String) : String
}
```

在这个接口里，fmt 是格式化字符串参数，它决定了格式化返回的结果。

### 19.1.1 整型、浮点型类型

对于整型、浮点型类型，格式化字符串的语法如下：

```
format_spec := [flags][width][.precision][specifier]
flags := '-' | '+' | '#' | '0'
width := integer
precision := integer
specifier := 'b' | 'B' | 'o' | 'O' | 'x' | 'X' | 'e' | 'E' | 'g' | 'G'
```

如上所示，格式化字符串中可以出现 flags（标志符）、width（宽度）、precision（精度）和 specifier（说明符）中的一种或者多种，下面说明各部分的使用方式。

1. width

为了方便演示，首先介绍 width。width 是正整数，表示格式化后字符串的宽度，如果宽度数值小于数值本身的长度,则格式化后的字符串不会被截断;默认字符串是右对齐的,如果宽度数值大于数值本身的长度,则字符串前面会被空白字符填充,示例代码如下:

```
//Chapter19/format_width_demo.cj

from std import format.Formatter

main() {
    var count = 10000

    //输出宽度为 8 的字符串
    println(count.format("8"))

    //输出宽度为 3 的字符串
    println(count.format("3"))

    count = -10000

    //输出宽度为 8 的字符串,包括前面的负号
    println(count.format("8"))
}
```

编译后运行该示例,输出如下:

```
   10000
10000
  -10000
```

format 库在 std 模块的 format 包里,需要先导入该包。因为变量 count 的长度为 5 位,第 1 个格式化字符串要求输出 8 位,所以第 1 个输出前面有 3 个空白字符;第 2 个格式化字符串要求输出 3 位,小于变量数值本身的长度,所以直接输出变量,没有截断;第 3 个格式化字符串也要求输出 8 位,但是因为是负数,符号也占用一位,所以前面有两个空白字符。

2. flags

flags 包括 4 个标识符,分别是'-'、'+'、'#'和'0',在一个格式化字符串里最多只能出现其中的一个标识符,如果包括多个标识符,则程序执行时会出现异常。

1) -标识符

表示左对齐,默认情况下数值类型是右对齐的,如果数值长度小于需要输出的宽度时,右对齐时在数值前面补足填充的字符,左对齐时在数值后面补足填充的字符。

2) +标识

数值为非负数时会在前面打印'+'符号,如果数值为负数,则忽略。

3) #标识符

打印数值的进制标识符。对于二进制,打印时会在数值前添加"0b"或者"0B";对于八进制,打印时会在数值前添加"0o"或者"0O";对于十六进制,打印时会在数值前添加"0x"

或者"0X"。本标识符需要配合 specifier(说明符)使用。

4) 0 标识符

对于数值长度小于需要输出的宽度时,使用该标识符表示不足的部分使用 0 补充。

flags 部分的示例代码如下(本示例不包括 # 标识符的示例,后续讲解 specifier 时会演示该标识符的用法):

```
//Chapter19/format_flags_demo.cj

from std import format.Formatter

main() {
    let count = 1000

    //左对齐输出 count 的值,会在后面补足填充的字符
    println(count.format("-8"))

    //默认右对齐输出 count 的值,会在前面补足填充的字符
    println(count.format("8"))

    //对于正数打印 + 符号
    println(count.format("+"))

    //0 也打印 + 符号
    println((0).format("+"))

    //对于负数忽略 + 符号
    println((-count).format("+"))

    //不足 8 位的部分使用 0 填充
    println(count.format("08"))
}
```

编译后运行该示例,输出如下:

```
1000
    1000
+1000
+0
-1000
00001000
```

3. precision

精度,对于整数类型,如果数值长度小于精度数字,则会在前面补足 0;对于浮点数类型,精度表示小数点后的位数,默认情况下精度为 6 位,示例代码如下:

```
//Chapter19/format_precision_demo.cj

from std import format.Formatter
```

```
main() {
    //整数长度小于宽度数值时,默认前面补充空格
    println((1000).format("8"))

    //整数长度小于精度数值时,默认前面补充 0
    println((1000).format(".8"))

    //精度表示浮点数小数点后的长度
    println((3.14).format(".8"))

    //默认精度是 6 位
    println(3.14)
}
```

编译后运行该示例,输出如下:

```
    1000
00001000
3.14000000
3.140000
```

### 4. specifier

specifier 表示说明符,针对整数类型,可选标识如下。

(1) b:以二进制形式输出,配合 # 标识符会在数值前添加 0b。
(2) B:以二进制形式输出,配合 # 标识符会在数值前添加 0B。
(3) o:以八进制形式输出,配合 # 标识符会在数值前添加 0o。
(4) O:以八进制形式输出,配合 # 标识符会在数值前添加 0O。
(5) x:以十六进制形式输出,配合 # 标识符会在数值前添加 0x。
(6) X:以十六进制形式输出,配合 # 标识符会在数值前添加 0X。

整数类型说明符的示例代码如下:

```
//Chapter19/int_specifier_demo.cj

from std import format.Formatter

main() {
    var number = 100

    //以二进制表示
    println(number.format("b"))

    //配合 # 标识符的八进制表示
    println(number.format("#o"))

    //配合 # 标识符的十六进制表示,宽度为 10 位
    println(number.format("#10X"))
}
```

编译后运行该示例,输出如下:

```
1100100
0o144
    0X64
```

在介绍浮点数类型标识符以前,先了解一下科学记数法。科学记数法是一种记数的方法,当需要标记或运算某个较大或较小且位数较多的数字时,用科学记数法可以避免浪费很多空间和时间。科学记数法把一个数表示成 a 与 10 的 n 次幂相乘的形式:

$$a \times 10^n$$

其中,$1 \leq |a| < 10$ 且不为分数形式,n 为整数。

针对浮点数类型,可选标识如下。

(1) e:科学记数法输出,使用小写的 e 表示 10 的幂。
(2) E:科学记数法输出,使用大写的 E 表示 10 的幂。
(3) g:精简表示,根据输出的长度自动选择十进制或者科学记数法,如果选择科学记数法,则使用小写的 e 表示 10 的幂。
(4) G:精简表示,根据输出的长度自动选择十进制或者科学记数法,如果选择科学记数法,则使用大写的 E 表示 10 的幂。

浮点类型说明符的示例代码如下:

```
//Chapter19/float_specifier_demo.cj

from std import format.Formatter

main() {
    var number1 = -9876543.21
    var number2 = 0.000123456

    //科学记数法表示
    println(number1.format("e"))
    println(number2.format("E"))

    var number3 = 123456.78

    //精简表示,这里自动按照十进制表示
    println(number3.format("g"))

    number3 = 1234567.89

    //精简表示,这里自动按照科学记数法表示
    println(number3.format("g"))
}
```

编译后运行该示例,输出如下:

```
-9.876543e+06
1.234560E-04
123457
1.23457e+06
```

### 19.1.2 字符类型

对于 char 类型，格式化语法如下：

```
format_spec := [flags][width]
flag := '-'
width := integer
```

其中，'-'标识符表示左对齐，width 表示字符串宽度，它们的详细用法和 19.1.1 节类似，示例代码如下：

```
//Chapter19/format_char_demo.cj

from std import format.Formatter

main() {
    let ch = 'a'

    //直接输出字符串
    println(ch)

    //输出宽度为 5 的字符串,默认右对齐,前面补足空格
    println(ch.format("5"))

    //输出宽度为 5 的字符串,左对齐,后面补足空格
    println(ch.format("-5"))
}
```

编译后运行该示例，输出如下：

```
a
    a
a
```

## 19.2 Console 类

Console 类提供从控制台读取和向控制台输出的功能，位于 std 模块的 console 包下，导入方式如下：

```
from std import console.*
```

控制台默认使用 UTF-8 编码，Windows 环境下需手动执行如下命令，将 windows 终端更改为 UTF-8 编码：

```
chcp 65001
```

Console 类包括 3 个主要的成员属性，如表 19-1 所示。

表 19-1 Console 类的成员

| 成 员 | 说 明 |
|---|---|
| public static prop stdIn：ConsoleReader | 标准输入的获取接口 |
| public static prop stdOut：ConsoleWriter | 标准输出的获取接口 |
| public static prop stdErr：ConsoleWriter | 标准错误的获取接口 |

上述属性的类型包括 ConsoleReader 和 ConsoleWriter，下面分别进行说明它们的主要用法。

### 19.2.1 ConsoleReader

提供从控制台读出数据并转换成字符或字符串的功能，读操作是同步的，内部设有缓冲区来保存控制台输入的内容，当到达控制台输入流的结尾时，读取函数将返回 None，例如在 UNIX 系统下按快捷键 Ctrl＋D 或在 Windows 系统下按快捷键 Ctrl＋Z。

ConsoleReader 的主要成员函数如下所示。

1) public func read()：Option＜Char＞

从控制台接收一个输入字符，以 Option＜Char＞的形式返回，如果读取到字符，返回 Option＜Char＞，否则返回 Option＜Char＞.None。在控制台输入字符时需要注意，输入时不会提交字符，只有按 Enter 键后才会提交，并且提交的字符不包括最后的换行符，示例代码如下：

```
//Chapter19/read_demo.cj

from std import console.*

main() {
    println("The characters you entered will be returned as is. Enter q to exit:")
    try {
        //读取一个字符
        var ch = Console.stdIn.read().getOrThrow()

        //判断输入的是否是字符 q,如果是就退出,否则继续循环
        while (ch != 'q') {
            //输入的字符是不是换行,如果不是就打印该字符
            if (ch != '\n') {
                println(ch)
            }

            //重新读取一个字符
            ch = Console.stdIn.read().getOrThrow()
        }
```

```
        } catch (e: Exception) {
            println("An error occurred:" + e.message)
        }
    }
```

编译后运行该示例,依次输入:

```
C(按 Enter 键)
J(按 Enter 键)
cj(按 Enter 键)
q(按 Enter 键)
```

输出如下:

```
The characters you entered will be returned as is. Enter q to exit:
C
C
J
J
cj
c
j
q
```

2) public func readln(): Option<String>

从控制台接收一行输入,以 Option<String>的形式返回,如果读取到字符,返回 Option<String>,否则返回 Option<String>.None,输入的信息不包含最后的换行符,示例代码如下:

```
//Chapter19/read_line_demo.cj

from std import console.*

main() {
    println("The characters you entered will be returned as is. Enter q to exit:")
    try {
        //读取一行字符
        var input = Console.stdIn.readln().getOrThrow()

        //判断输入的是否是字符 q,如果是就退出,否则继续循环
        while (input != "q") {
            println(input)

            //重新读取一行
            input = Console.stdIn.readln().getOrThrow()
        }
    } catch (e: Exception) {
        println("An error occurred:" + e.message)
    }
}
```

编译后运行该示例，依次输入：

```
Hello(按 Enter 键)
cangjie(按 Enter 键)
q(按 Enter 键)
```

输出如下：

```
Hello
Hello
cangjie
cangjie
q
```

3) public func readToEnd()：Option < String >

从控制台读取所有字符，以 Option < String > 的形式返回，如果读取到字符，返回 Option < String >，否则返回 Option < String >.None，读取失败时会返回 Option < String >.None。

4) public func readUntil(ch：Char)：Option < String >

从控制台读取数据直到读取到字符 ch 为止，ch 包含在结果中；如果读取到文件结束符 EOF，将返回读取到的所有信息；读取失败时会返回 Option < String >.None。

5) public func readUntil(predicate：(Char) —> Bool)：Option < String >

从控制台读取数据直到读取到的字符满足 predicate 条件时结束，满足 predicate：(Char) —> Bool 条件的字符也包含在结果中；如果读取到文件结束符 EOF，将返回读取到的所有信息；读取失败时会返回 Option < String >.None，示例代码如下：

```
//Chapter19/read_until_demo.cj

from std import console.*

var lastChr = Option < Char >.None

func untilDouble9(newChr: Char): Bool {
    //如果上一个字符存在
    if (let Some(last) <- lastChr) {
        //如果上一个字符和现在的字符都是 9
        if (last == '9' && newChr == '9') {
            return true
        } else {
            lastChr = newChr
            return false
        }
    } else {
        lastChr = newChr
    }

    return false
```

```
}

main() {
    println("Input string, including two consecutive characters 9:")
    try {
        //读取字符串,如果读取的内容包括两个连续的9,则返回截止这两个9的子字符串
        var input = Console.stdIn.readUntil(untilDouble9)

        //判断结果是不是None,是就继续循环
        while (let None <- input) {
            //重新读取
            input = Console.stdIn.readUntil(untilDouble9)
        }

        println(input.getOrThrow())
    } catch (e: Exception) {
        println("An error occurred:" + e.message)
    }
}
```

编译后运行该示例,依次输入:

```
cangjie is the best language,(按 Enter 键)
with a score of 99. Hurry up and learn it.(按 Enter 键)
```

输出如下:

```
Input string, including two consecutive characters 9:
cangjie is the best language,
with a score of 99. Hurry up and learn it.
cang jie is the best language,
with a score of 99
```

6) public func read(arr: Array<Byte>): Int64

从控制台读取并放入数组 arr 中,返回读取到的字节长度;该函数存在风险,可能读取出来的结果恰好把 UTF-8 编码的代码点(code point)从中截断。

## 19.2.2　ConsoleWriter

提供线程安全的输出数据到控制台的功能,每次 write 调用写到控制台的结果是完整的。

ConsoleWriter 的主要函数如下所示。

1) public func flush(): Unit

刷新输出流。

2) public func write(v: String): Unit

输出字符串参数 v 到控制台。

3) public func writeln(v: String): Unit

输出字符串参数 v 到控制台并换行。

write 和 writeln 还有多个重载函数,就不再一一列出了。

## 19.3 Random 类

Random 是随机数生成类,可以生成多种数据类型的随机数,位于 std 模块的 random 包下,导入方式如下:

```
from std import random.*
```

该类的主要函数如下。

1) public open init()

无参构造函数。

2) public open init(seed: UInt64)

使用给定的种子创建新的 Random 对象。Random 类生成的随机数并不是真正的"随机数",而是一种"伪随机数",在同一台计算机上,种子相同,生成的随机数的序列也是相同的,后面会给出示例。

3) public open mut prop seed: UInt64

设置或者获取随机数的种子大小,因为种子相同生成的随机数相同,所以,可以每次设置不同的种子来生成相对更"随机"的随机数。

4) public open func nextBool(): Bool

获取布尔类型的随机值。

5) public open func nextUInt8(): UInt8

获取 UInt 类型的随机值。

获取 Bool 和 UInt 类型的随机数,示例代码如下:

```
//Chapter19/random_demo.cj

from std import random.Random

main() {
    //生成随机数对象
    let random = Random()

    //生成 Bool 类型的随机数
    println(random.nextBool())

    //生成 UInt8 类型的随机数
    println(random.nextUInt8())
}
```

编译后运行该示例,可能的输出如下:

```
true
110
```

6) public open func nextUInt8(upper: UInt8): UInt8

获取一个 UInt8 类型的随机数,该随机数大于或等于 0,但小于 upper。下面给出一个调用该函数的示例,该示例两次都设置相同的种子,每次随机生成 5 个数字,代码如下:

```
//Chapter19/pseudo_random_demo.cj

from std import random.Random

main() {
    //生成随机数对象
    let random = Random()

    //设置随机数对象的种子为100
    random.seed = 100

    //设置随机数数量
    var count = 5
    println("The first random number sequence:")
    while (count > 0) {
        //生成不大于 20 的 UInt8 类型的随机数
        println(random.nextUInt8(20))
        count--
    }

    //重新设置随机数对象的种子为100
    random.seed = 100

    //设置随机数数量
    count = 5
    println("The second random number sequence:")
    while (count > 0) {
        //生成不大于 20 的 UInt8 类型的随机数
        println(random.nextUInt8(20))
        count--
    }
}
```

编译后运行该示例,可能的输出如下:

```
The first random number sequence:
15
14
5
2
6
The second random number sequence:
15
```

```
14
5
2
6
```

可以看到,在相同的种子下,生成的随机数的序列是相同的。

7) public open func nextFloat16(): Float16

获取一个 Float16 类型的随机数,该随机数大于或等于 0,但小于 1,示例代码如下:

```
//Chapter19/float_random_demo.cj

from std import random.Random

main() {
    //生成随机数对象
    let random = Random()

    //设置随机数的数量
    var count = 5
    while (count > 0) {
        //生成 Float16 类型的随机数
        println(random.nextFloat16())
        count --
    }
}
```

编译后运行该示例,可能的输出如下:

```
0.510254
0.745117
0.434814
0.242920
0.000456
```

除了上述介绍的函数,Random 还有大量的产生其他数据类型随机数的函数,另外几个随机数函数如下。

- public open func nextInt32(): Int32
- public open func nextUInt32(upper: UInt32): UInt32
- public open func nextFloat64(): Float64

这些函数的调用方式和本节介绍的其他函数类似,可以参考相应的 API 文档,此处就不再赘述。

## 19.4　数学库

数学库提供常见的通用数学运算、常数定义、浮点数处理等,位于 std 模块的 math 包下,导入方式如下:

```
from std import math.*
```

下面按照常数、函数的类别,分别介绍主要的几个常量和函数。

### 19.4.1 常数

#### 1. 双精度常数

双精度浮点数的类型为 Float64,扩展出了如表 19-2 所示的双精度常数及相关的判断函数。

表 19-2　Float64 扩展的常数

| 成　　员 | 说　　明 |
| --- | --- |
| public static prop NaN：Float64 | 双精度浮点数的非数 |
| public static prop Inf：Float64 | 双精度浮点数的无穷数 |
| public static prop PI：Float64 | 双精度浮点数的圆周率常数 |
| public static prop E：Float64 | 双精度浮点数的自然常数 |
| public static prop Max：Float64 | 双精度浮点数的最大值 |
| public static prop Min：Float64 | 双精度浮点数的最小值 |
| public static prop MinDenormal：Float64 | 双精度浮点数的最小次正规数 |
| public static prop MinNormal：Float64 | 双精度浮点数的最小正规数 |
| public func isInf()：Bool | 判断某个浮点数是否为无穷数值 |
| public func isNaN()：Bool | 判断某个浮点数是否为非数值 |
| public func isNormal()：Bool | 判断某个浮点数是否为常规数值 |

#### 2. 单精度常数

单精度浮点数的类型为 Float32,扩展出了如表 19-3 所示的单精度常数及相关的判断函数。

表 19-3　Float32 扩展的常数

| 成　　员 | 说　　明 |
| --- | --- |
| public static prop NaN：Float32 | 单精度浮点数的非数 |
| public static prop Inf：Float32 | 单精度浮点数的无穷数 |
| public static prop PI：Float32 | 单精度浮点数的圆周率常数 |
| public static prop E：Float32 | 单精度浮点数的自然常数 |
| public static prop Max：Float32 | 单精度浮点数的最大值 |
| public static prop Min：Float32 | 单精度浮点数的最小值 |
| public static prop MinDenormal：Float32 | 单精度浮点数的最小次正规数 |
| public static prop MinNormal：Float32 | 单精度浮点数的最小正规数 |
| public func isInf()：Bool | 判断某个浮点数是否为无穷数值 |
| public func isNaN()：Bool | 判断某个浮点数是否为非数值 |
| public func isNormal()：Bool | 判断某个浮点数是否为常规数值 |

#### 3. 半精度常数

半精度浮点数的类型为 Float16,扩展出了如表 19-4 所示的半精度常数及相关的判断函数。

表 19-4　Float16 扩展的常数

| 成　员 | 说　明 |
| --- | --- |
| public static prop NaN：Float16 | 半精度浮点数的非数 |
| public static prop Inf：Float16 | 半精度浮点数的无穷数 |
| public static prop PI：Float16 | 半精度浮点数的圆周率常数 |
| public static prop E：Float16 | 半精度浮点数的自然常数 |
| public static prop Max：Float16 | 半精度浮点数的最大值 |
| public static prop Min：Float16 | 半精度浮点数的最小值 |
| public static prop MinDenormal：Float16 | 半精度浮点数的最小次正规数 |
| public static prop MinNormal：Float16 | 半精度浮点数的最小正规数 |
| public func isInf()：Bool | 判断某个浮点数是否为无穷数值 |
| public func isNaN()：Bool | 判断某个浮点数是否为非数值 |
| public func isNormal()：Bool | 判断某个浮点数是否为常规数值 |

### 4. 64 位有符号整数

64 位有符号整数的类型为 Int64，扩展出了如表 19-5 所示的最大值和最小值的常数。

表 19-5　Int64 扩展的常数

| 成　员 | 说　明 |
| --- | --- |
| public static prop Max：Int64 | 64 位有符号整数的最大值 |
| public static prop Min：Int64 | 64 位有符号整数的最小值 |

除了 64 位有符号整数类型 Int64，类似的整数类型还有 32 位有符号整数类型 Int32、16 位有符号整数类型 Int16、8 位有符号整数类型 Int8、64 位无符号整数类型 UInt64、32 位无符号整数类型 UInt32、16 位无符号整数类型 UInt16、8 位无符号整数类型 UInt8，这些类型的常数是类似的，不再一一列出。

常数的示例代码如下：

```
//Chapter19/constant_demo.cj

from std import math. *

main() {
    //双精度自然常数
    println("E64: ${Float64.E}")

    //单精度无穷数
    println("Inf32: ${Float32.Inf}")

    //半精度最大浮点数
    println("MaxFloat16: ${Float16.Max}")

    //64 位最大整数
    println("MaxInt64: ${Int64.Max}")
```

```
    //双精度最小浮点数
    println("MinFloat64:${Float64.Min}")

    //双精度非数
    println("NaN64:${Float64.NaN}")

    //双精度圆周率
    println("PI64:${Float64.PI}")
}
```

编译后运行该示例,输出如下:

```
E64:2.718282
Inf32:inf
MaxFloat16:65504.000000
MaxInt64:9223372036854775807
MinFloat64:-17976931348623157081452742373170435679807056752584499659891747680315726078002853876058955863276687817154045895351438246423432132688946418276846754670353751698604991057655128207624549009038932894407586850845513394230458323690322294816580855933212334827479782620414472316873817771809192998812504040261841248583568.000000
NaN64:nan
PI64:3.141593
```

## 19.4.2 函数

**1. 求绝对值**

求整数或者浮点数的绝对值。

- public func abs(x:Int8):Int8
- public func abs(x:Int16):Int16
- public func abs(x:Int32):Int32
- public func abs(x:Int64):Int64
- public func abs(x:Float64):Float64
- public func abs(x:Float32):Float32
- public func abs(x:Float16):Float16

**2. 浮点数向上取整**

获取大于或等于参数的最小整值。

- public func ceil(x:float16):float16
- public func ceil(x:float32):float32
- public func ceil(x:float64):float64

**3. 浮点数向下取整**

获取小于或等于参数的最大整值。

- public func floor(x:float16):float16

- public func floor(x：float32)：float32
- public func floor(x：float64)：float64

### 4. 浮点数截断取整

获取截断后整数部分的浮点数。

- public func trunc(x：float16)：float16
- public func trunc(x：float32)：float32
- public func trunc(x：float64)：float64

### 5. 自然常数 e 的幂

获取 e 的 a 次幂。

- public func exp(a：float16)：float16
- public func exp(a：float32)：float32
- public func exp(a：float64)：float64

### 6. 2 的幂

获取 2 的 a 次幂。

- public func exp2(a：float16)：float16
- public func exp2(a：float32)：float32
- public func exp2(a：float64)：float64

### 7. 自然常数 e 的对数

获取以自然常数 e 为底的 x 的对数。

- public func log(x：float16)：float16
- public func log(x：float32)：float32
- public func log(x：float64)：float64

### 8. 2 的对数

获取以 2 为底的 x 的对数。

- public func log2(x：float16)：float16
- public func log2(x：float32)：float32
- public func log2(x：float64)：float64

### 9. 正弦值

获取浮点数 x 的正弦值。

- public func sin(x：Float64)：Float64
- public func sin(x：Float32)：Float32
- public func sin(x：Float16)：Float16

### 10. 余弦值

获取浮点数 x 的余弦值。

- public func cos(x：Float64)：Float64
- public func cos(x：Float32)：Float32

- public func cos(x：Float16)：Float16

### 11．正切值

获取浮点数 x 的正切值。

- public func tan(x：Float64)：Float64
- public func tan(x：Float32)：Float32
- public func tan(x：Float16)：Float16

数学函数的示例代码如下：

```
//Chapter19/math_demo.cj

from std import math.*

main() {
    let num1 = 3.14
    let num2 = -3.14

    //向上取整(正数)
    println(ceil(num1))

    //向上取整(负数)
    println(ceil(num2))

    //向下取整(正数)
    println(floor(num1))

    //向下取整(负数)
    println(floor(num2))

    //截断取整(正数)
    println(trunc(num1))

    //截断取整(负数)
    println(trunc(num2))

    let num = 10.0

    //自然常数的 10 次幂
    println(exp(num))

    //2 的 10 次幂
    println(exp2(num))

    let value = 512.0

    //以自然常数为底的 512 的对数
    println(log(value))

    //以 2 为底的 512 的对数
```

```
    println(log2(value))

    //π 的 1/6
    let rad = Float64.PI / 6.0

    //取正弦值
    println(sin(rad))

    //取余弦值
    println(cos(rad))

    //取正切值
    println(tan(rad))
}
```

编译后运行该示例,输出如下:

```
4.000000
-3.000000
3.000000
-4.000000
3.000000
-3.000000
22026.465795
1024.000000
6.238325
9.000000
0.500000
0.866025
0.577350
```

## 19.5 转换库

转换库提供从字符串转换到特定类型的转换函数,该库位于 std 模块的 convert 包下,导入方式如下:

```
from std import convert.*
```

在仓颉语言中提供字符串到特定类型转换的泛型接口是 Parsable,定义如下:

```
public interface Parsable<T> {
    //从字符串中解析特定类型,参数 value 为待解析的字符串
    static func parse(value: String): T

    //从字符串中解析特定类型,参数 value 为待解析的字符串,如果转换失败,则返回 Option<T>.None
    static func tryParse(value: String): Option<T>
}
```

仓颉转换库为基础数据类型实现该接口的扩展,从而为基础类型添加转换能力。

典型基础类型的转换函数如下所示。

### 1. Bool

- public static func parse(data：String)：Bool

将 Bool 类型字面量的字符串 data 转换为 Bool 值，当字符串为空或转换失败时抛出 IllegalArgumentException 异常。

- public static func tryParse(data：String)：Option＜Bool＞

将 Bool 类型字面量的字符串 data 转换为 Option＜Bool＞值，转换失败返回 Option＜Bool＞.None。

### 2. Char

- public static func parse(data：String)：Char

将 Char 类型字面量的字符串 data 转换为 Char 值，当字符串为空或转换失败时，或字符串中含有无效的 UTF-8 字符时抛出 IllegalArgumentException 异常。

- public static func tryParse(data：String)：Option＜Char＞

将 Char 类型字面量的字符串 data 转换为 Option＜Char＞值，转换失败返回 Option＜Char＞.None，当字符串中含有无效的 UTF-8 字符时抛出 IllegalArgumentException 异常。

### 3. Int8

- public static func parse(data：String)：Int8

将 Int8 类型字面量的字符串 data 转换为 Int8 值，当字符串为空，首位为"+"，转换失败，或转换后超出 Int8 范围，或字符串中含有无效的 UTF-8 字符时，抛出 IllegalArgumentException 异常。

- public static func tryParse(data：String)：Option＜Int8＞

将 Int8 类型字面量的字符串 data 转换为 Option＜Int8＞值，转换失败返回 Option＜Int8＞.None，当字符串中含有无效的 UTF-8 字符时抛出 IllegalArgumentException 异常。

其他基础类型，如 Int16、Int32、Int64、UInt8、UInt16、UInt32、UInt64、Float16、Float32、Float64 等，也实现了 Parsable 接口，使用方式和上述类型类似，不再一一列举。

字符串转换为其他类型的示例代码如下：

```
//Chapter19/string_to_other_type_demo.cj

from std import convert.*

main() {
    var strValue = "true"
    let boolValue = Bool.tryParse(strValue)
    match (boolValue) {
        case Some(value) => println("The value is ${value}")
        case None => println("The value is not boolean type!")
    }

    strValue = "3.14"
    let floatValue = Float64.tryParse(strValue)
```

```
    match (floatValue) {
        case Some(value) => println("The value is ${value}")
        case None => println("The value is not float type!")
    }

    strValue = "100"
    let intValue = Int32.tryParse(strValue)
    match (intValue) {
        case Some(value) => println("The value is ${value}")
        case None => println("The value is not integer type!")
    }
}
```

编译后运行该示例,输出如下:

```
The value is true
The value is 3.140000
The value is 100
```

## 19.6　base64 包

base64 包提供了字符串的 Base64 编码及解码能力,位于 encoding 模块的 base64 包下,导入方式如下:

```
from encoding import base64.*
```

该包拥有以下两个函数,可以把 8 位无符号数数组与 Base64 编码字符串互相转换。
- public func toBase64String(data：Array < UInt8 >)：String
- public func fromBase64String(data：String)：Option < Array < UInt8 >>

二进制数据不适合在 HTTP 环境下直接传输,可以通过 Base64 编码的方式把这些数据转换为基于 64 个可打印字符形成的字符串,这样就可以用于 HTTP 传输了。在实际应用中,可以把字符串先转换为 ASCII 码表示的 8 位无符号数数组,然后对这个数组进行 Base64 编码,把编码后的 Base64 字符串用于网络传输,在接收端接收到 Base64 字符串后,再对其解码,得到 8 位无符号数数组,再把这个数组转换为字符串即可,示例代码如下:

```
//Chapter19/base64_demo.cj

from encoding import base64.*

main() {
    //定义 8 位无符号数组
    let ByteArray = Array<UInt8>([104, 101, 108, 108, 111, 32, 99, 97, 110, 103, 106, 105, 101, 33])

    //将数组转换为 base64 字符串形式
```

```
        let base64Str = toBase64String(ByteArray)

        //打印 base64 字符串
        println(base64Str)

        //将 base64 字符串转换为数组
        let convertResult = fromBase64String(base64Str)

        match (convertResult) {
            case Some(value) =>
                //把数组的每个值转换为字符打印
                for (num in value) {
                    print(chr(num))
                }
            case None => println("Invalid conversion!")
        }
}
```

编译后运行该示例,输出如下:

```
aGVsbG8gY2FuZ2ppZSE=
hello cangjie!
```

## 19.7　hex 包

hex 包提供了字符串的 Hex 编码及解码能力,位于 encoding 模块的 hex 包下,导入方式如下:

```
from encoding import hex.*
```

该包拥有以下两个函数,可以把 8 位无符号数数组与 Hex 编码字符串互相转换。
- public func fromHexString(data: String): Option<Array<UInt8>>
- public func toHexString(data: Array<UInt8>): String

一个 8 位的字节数据,分成高低各 4 位,因为 4 位数据的取值范围为[0~15],这两个 4 位的数据可以分别转换为一个十六进制的数字,这样就可以用两个十六进制的数字表示一个 8 位字节数据,再把这两个十六进制数字转换为字符串,就形成了 Hex 编码字符串。在网络传输和串口调试等场景下,Hex 编码有广泛的应用,示例代码如下:

```
//Chapter19/hex_demo.cj

from encoding import hex.*

func main() {
    //定义 8 位无符号数组
    let ByteArray = Array<UInt8>([104, 101, 108, 108, 111, 32, 99, 97, 110, 103, 106, 105, 101, 33])
```

```
        //打印数组数据,使用逗号分隔
        var index = 0
        while (index < byteArray.size) {
            print(byteArray[index])
            if (index < byteArray.size - 1) {
                print(",")
            }

            index++
        }

        //打印空行
        println("")

        //将数组转换为十六进制字符串
        let hexStr = toHexString(ByteArray)

        //打印十六进制字符串
        index = 0
        while (index < hexStr.size) {
            //为了便于观察,每两个字符使用逗号分隔
            print("${hexStr[index]}${hexStr[index + 1]}")
            if (index < hexStr.size - 2) {
                print(",")
            }

            index += 2
        }

        //打印空行
        println("")

        //将十六进制字符串转换为数组
        let array = fromHexString(hexStr).getOrThrow()

        //把数组的每个值转换为字符后打印
        var indx = 0
        while (indx < array.size) {
            print(chr(UInt32(array[indx])))
            indx++
        }
}
```

编译后运行该示例,输出如下:

```
104,101,108,108,111,32,99,97,110,103,106,105,101,33
68,65,6c,6c,6f,20,63,61,6e,67,6a,69,65,21
hello cangjie!
```

## 19.8　时间库

时间库提供了与时间操作相关的能力，包括时间的读取、时间计算、基于时区的时间转换、时间的序列化和反序列化等，位于 std 模块的 time 包下，导入方式如下：

```
from std import time.*
```

下面介绍相关的类和枚举的基本用法。

### 19.8.1　Month 枚举

表示月份的枚举，包括以下所示的 12 个构造器，分别代表 1 月份到 12 月份：
- January
- February
- March
- April
- May
- June
- July
- August
- September
- October
- November
- December

Month 类包括的主要函数如下。
- public static func of(mon: Int64): Month

把数字形式的参数转换为对应的 Month 实例，参数取值范围为 1～12，表示一年的 12 个月，当 mon 不在 [1, 12] 范围内时，抛出 IllegalArgumentException 异常。
- public func value(): Int64

将 Month 实例转换为 Int64 类型的月份数字。
- public func toString(): String

返回 Month 实例的月份字符串。
- public operator func +(n: Int64): Month

重载加法操作符，可以对一个 Month 实例加上一个整数，表示一个新的 Month，如果新的 Month 月份大于 12，则会进行取模运算，保证结果为合法的月份。
- public operator func -(n: Int64): Month

重载减法运算符，可以对一个 Month 实例减去一个整数，表示一个新的 Month，如果新的 Month 月份小于 1，则会进行取模运算，保证结果为合法的月份。
- public operator func ==(r: Month): Bool

重载判等操作符，判断两个 Month 的月份是否相等。
- public operator func !=(r: Month): Bool

重载判不等操作符，判断两个 Month 的月份是否不相等。

Month 类型的示例代码如下所示，该示例代码将打印出 12 个月份以及每个月份前后月份的信息：

```
//Chapter19/month_demo.cj

from std import time.*

main() {
    var month = 1

    //循环获取 12 个月份的信息
    while (month <= 12) {
        //获取 month 对应的 Month 对象
        let currentMonth = Month.of(month)

        //上一个月份
        var preMonth: Month
        if (month == 1) {
            preMonth = Month.December
        } else {
            preMonth = currentMonth - 1
        }

        //下一个月份
        let nextMonth = currentMonth + 1

        //计算月份序号
        var ordStr = match (currentMonth.value()) {
            case 1 => "1st"
            case 2 => "2nd"
            case 3 => "3rd"
            case _ => currentMonth.value().toString() + "th"
        }

        //输出月份及前后月份信息
        println(
            "The ${ordStr} month is ${currentMonth.toString()}," +
                "the previous month was ${preMonth.toString()}, and the next month is ${nextMonth.toString()}"
        )

        month++
    }
}
```

编译后运行该示例，输出如下：

```
The 1st month is January, the previous month was December, and the next month is February
The 2nd month is February, the previous month was January, and the next month is March
The 3rd month is March, the previous month was February, and the next month is April
The 4th month is April, the previous month was March, and the next month is May
The 5th month is May, the previous month was April, and the next month is June
The 6th month is June, the previous month was May, and the next month is July
The 7th month is July, the previous month was June, and the next month is August
The 8th month is August, the previous month was July, and the next month is September
The 9th month is September, the previous month was August, and the next month is October
The 10th month is October, the previous month was September, and the next month is November
The 11th month is November, the previous month was October, and the next month is December
The 12th month is December, the previous month was November, and the next month is January
```

## 19.8.2　DayOfWeek 枚举

表示一周中某一天的枚举，包括以下 7 个构造器，分别代表周日、周一到周六：

- Sunday
- Monday
- Tuesday
- Wednesday
- Thursday
- Friday
- Saturday

DayOfWeek 包括的函数和 Month 一致，这里不再赘述，需要注意的是数值和 Weekday 转换的问题，数字 0 表示的是周日，1～6 分别表示周一到周六，示例代码如下：

```
//Chapter19/weekday_demo.cj

from std import convert.*
from std import console.*
from std import time.*

main() {
    println("Please enter a number and press enter, we will display the corresponding day of the week.")
    println("Press q to exit.")
    var inPut = Console.stdIn.read().getOrThrow()

    //判断输入字符是不是 q
    while (inPut != 'q') {
        let inputNum = Int64.tryParse(inPut.toString())
        try {
            let weekNum = inputNum.getOrThrow()
            let weekDay = DayOfWeek.of(weekNum)
```

```
                println(weekDay)
            } catch (_) {} finally {
                inPut = Console.stdIn.read().getOrThrow()
            }
        }
    }
```

编译后运行该示例,分别输入 1、2、3、q,输出如下:

```
Please enter a number and press enter,we will display the corresponding day of the week.
Press q to exit.
1
Monday
2
Tuesday
3
Wednesday
q
```

### 19.8.3 Duration 类

Duration 表示一段时间间隔,精度为 ns,可以表示 $-2^{63}$ 到 $2^{63}-1$ 纳秒范围;Duration 的每个时间单位(小时、分、秒等)均用整数表示,如果实际值不为整数,则向绝对值小的方向取整(例如,2.8h 取整为 2h,$-$2.8h 取整为$-$2h,0.5min 取整为 0min,$-$0.7s 取整为 0s)。

主要成员如下所示。

**1. 表示 Duration 实例的成员属性**(见表 19-6)

表 19-6 Durations 成员属性

| 成 员 属 性 | 说　　明 |
| --- | --- |
| public static prop nanosecond:Duration | 1ns 时间间隔的 Duration 实例 |
| public static prop microsecond:Duration | 1μs 时间间隔的 Duration 实例 |
| public static prop millisecond:Duration | 1ms 时间间隔的 Duration 实例 |
| public static prop second:Duration | 1s 时间间隔的 Duration 实例 |
| public static prop minute:Duration | 1min 时间间隔的 Duration 实例 |
| public static prop hour:Duration | 1h 时间间隔的 Duration 实例 |
| public static prop day:Duration | 1 天时间间隔的 Duration 实例 |
| public static prop Zero:Duration | 0ns 时间间隔的 Duration 实例 |
| public static prop Max:Duration | $2^{63}-1$ns 时间间隔的 Duration 实例 |
| public static prop Min:Duration | $-2^{63}$ns 时间间隔的 Duration 实例 |

**2. 返回 Duration 实例在不同单位下的大小**

- public func toDays():Int64
- public func toHours():Int64

- public func toMinutes()：Int64
- public func toSeconds()：Int64
- public func toMilliseconds()：Int64
- public func toMicroseconds()：Int64
- public func toNanoseconds()：Int64

返回 Duration 实例以函数名称为单位的大小。

3．返回 Duration 实例的字符串表示形式

- public func toString()：String

返回形如 1d2h3m4s5ms6us7ns 格式的字符串,表示"1 天 2 小时 3 分钟 4 秒 5 毫秒 6 微秒 7 纳秒",如果某个单位的值为 0,则省略该单位;如果全为 0,则返回 0s。

4．操作符重载

- public operator func +(r：Duration)：Duration
- public operator func -(r：Duration)：Duration
- public operator func *(r：Int64)：Duration
- public operator func *(r：Float64)：Duration
- public operator func /(r：Int64)：Duration

两个 Duration 实例会把时间间隔相加,得到新的 Duration 实例;两个 Duration 实例相减,用第 1 个实例的时间间隔减去第 2 个实例的时间间隔,两者的差形成一个新的 Duration 实例。Duration 实例乘以一个整数会把时间间隔乘以该整数,并返回新的 Duration 实例;Duration 实例乘以一个浮点数会把时间间隔乘以该浮点数,并返回新的 Duration 实例,实例不足 1 纳秒的部分,向绝对值小的方向取整;Duration 实例除以一个整数会把时间间隔除以该整数,并返回新的 Duration 实例。

- public operator func ==(r：Duration)：Bool
- public operator func !=(r：Duration)：Bool
- public operator func >=(r：Duration)：Bool
- public operator func >(r：Duration)：Bool
- public operator func <=(r：Duration)：Bool
- public operator func <(r：Duration)：Bool

两个 Duration 实例比较,就是两个实例的时间间隔相比较。

5．其他成员函数

- public func abs()：Duration

返回一个新的 Duration 实例,其值大小为当前 Duration 实例的绝对值。

- public static func since(t：DateTime)：Duration

计算从参数 t 开始到当前时间为止的时间间隔。

- public static func until(t：DateTime)：Duration

计算从当前时间开始到参数 t 为止的时间间隔。

- public func compare(rhs: Duration): Ordering

比较当前 Duration 实例与参数 rhs 的关系,如果大于,则返回 Ordering.GT;如果等于,则返回 Ordering.EQ;如果小于,则返回 Ordering.LT。

Duration 函数的示例代码如下:

```
//Chapter19/duration_demo.cj

from std import random.*
from std import time.*

main() {
    let rand = Random()

    //随机生成两个 Duration 实例
    let duration1 = Duration.minute * rand.nextInt64(100) + Duration.second * rand.nextInt64(100) + Duration.millisecond * rand.nextInt64(100)
    let duration2 = Duration.hour * rand.nextInt64(100) + Duration.microsecond * rand.nextInt64(100) + Duration.nanosecond * rand.nextInt64(100)

    //打印两个 Duration 实例
    println("The first duration is ${duration1}")
    println("The second duration is ${duration2}")

    //计算 Duration 实例差值
    let diff = duration2 - duration1

    println("The difference between the first and second is ${diff}, which is equivalent to ${diff.toSeconds()} seconds")
}
```

编译后运行该示例,可能的输出如下:

```
The first duration is 44m33s58ms
The second duration is 1d12h48us66ns
The difference between the first and second is 1d11h15m26s942ms48us66ns, which is equivalent to 126926 seconds
```

### 19.8.4  TimeZone

表示具体的时区信息,可以自定义时区,也可以加载系统中已存在的时区,TimeZone 类的常用成员和函数如下所示。

**1. 构造函数**

- public init(id: String, offset: Duration)

使用参数时区 id 和偏移 offset 构造自定义 TimeZone 实例,其中,offset 精度为秒,相对于 UTC 时区向东为正。

2．属性
- public prop id：String

获取当前 TimeZone 实例所关联的时区 id。

3．加载时区函数
- public static func load(id：String)：TimeZone

从系统中加载参数 id 指定的时区，id 格式符合标准时区 id 规范，如 Asia/Shanghai 等。
- public static func loadFromPaths(id：String, tzpaths：Array<String>)：TimeZone

从参数 tzpaths 指定的路径中加载参数 id 指定的时区。
- public static func loadFromTZData(id：String, data：Array<UInt8>)：TimeZone

使用参数 id 指定的时区和参数 data 指定的时区数据构造一个自定义 TimeZone 实例，id 可以是任何合法字符串，data 需要满足 IANA 时区文件格式，加载成功时获得 TimeZone 实例，否则抛出异常。

4．返回字符串表示形式
- public func toString()：String

返回时区名称。

5．静态成员变量
- public static let Local：TimeZone
- public static let UTC：TimeZone

Local 和 UTC 都是 TimeZone 类的静态成员变量，Local 表示本地时区，UTC 本身指"世界协调时间"，这里表示时区偏移量为 0 的时区。

TimeZone 的示例代码如下：

```
//Chapter19/location_demo.cj

from std import time.*

main() {
    let loc = TimeZone.Local
    let utc = TimeZone.load("UTC")
    let locShanghai = TimeZone.load("Asia/Shanghai")

    println(loc)
    println(utc)
    println(locShanghai)
}
```

编译后运行该示例，输出如下：

```
localtime
UTC
Asia/Shanghai
```

### 19.8.5 DateTime

DateTim 是 struct 类型,提供了时间的读取和计算、基于时区的时间转换、序列化和反序列化等多种功能。

在介绍 DateTim 成员以前,需要先了解两个关于时间的基础知识,一个是 UNIX 时间,另一个是 RFC 3339 时间格式。

(1) UNIX 时间:一个数字,指的是从 UTC 时区的 1970 年 1 月 1 日 0 点开始计算,到现在的秒数和纳秒数。

(2) RFC 3339 时间格式:经 IETF(Internet Engineering Task Force)起草和审核的关于 Internet 时间的标准,它的时间格式定义语法如下:

date-fullyear=4 位数字。

date-month=2 位数字;01~12。

date-mday=2 位数字;01~28,01~29,01~30,01~31,和具体的年月有关。

time-hour=2 位数字;00~23。

time-minute=2 位数字;00~59。

time-second=2 位数字;00~58,00~59,00~60,和闰秒有关。

time-secfrac="."1 位或者多位数字。

time-numoffset=("+" / "−") time-hour ":" time-minute。

time-offset="Z" / time-numoffset。

partial-time=time-hour ":" time-minute ":" time-second[time-secfrac]。

full-date=date-fullyear "-" date-month "-" date-mday。

full-time=partial-time time-offset。

date-time=full-date "T" full-time。

**1. 成员属性**

DateTim 主要成员如表 19-7 所示。

表 19-7 DateTime 的属性成员

| 成 员 属 性 | 说　　明 |
| --- | --- |
| public static prop UnixEpoch:DateTime | 该属性为静态属性,表示 UNIX 时间纪元的开始时间,也就是 1970-01-01T00:00:00Z |
| public prop year:Int64 | 当前实例的年份 |
| public prop month:Month | 当前实例的月份 |
| public prop monthValue:Int64 | 当前实例的月,为整数类型,以时间 2023-10-01T08:43:25Z 为例,该属性将返回 10 |
| public prop dayOfMonth:Int64 | 当前实例以月算的天数,以时间 2023-10-01T08:43:25Z 为例,该属性将返回 1 |
| public prop dayOfWeek:DayOfWeek | 当前实例以周算的天数 |

| 成 员 属 性 | 说 明 |
| --- | --- |
| public prop dayOfYear: Int64 | 当前实例以年算的天数,以时间 2023-10-01T08:43:25Z 为例,该属性将返回 274 |
| public prop hour: Int64 | 当前实例的小时部分,以时间 2023-10-01T08:43:25Z 为例,该属性将返回 8 |
| public prop minute: Int64 | 当前实例的分钟部分,以时间 2023-10-01T08:43:25Z 为例,该属性将返回 43 |
| public prop second: Int64 | 当前实例的秒部分,以时间 2023-10-01T08:43:25Z 为例,该属性将返回 25 |
| public prop nanosecond: Int64 | 当前实例的纳秒部分,以时间 2023-10-01T08:43:25Z 为例,该属性将返回 0 |
| public prop isoWeek | 当前实例基于 ISO8601 标准的年份和基于年的周数返回值,以时间 2023-10-01T08:43:25Z 为例,该属性将返回 (2023,39) |
| public prop zone: TimeZone | 当前实例所关联的 TimeZone |
| public prop zoneId: String | 当前实例所关联的时区 id,以时间 2023-10-01T08:43:25Z 为例,该属性将返回 GMT+8 |
| public prop zoneOffset: Duration | 当前实例所关联的时间偏移,以时间 2023-10-01T08:43:25Z 为例,该属性将返回 8h 表示的时间间隔 |

**2. 构造实例的相关函数**

- public static func fromUnixTimeStamp(d: Duration): DateTime

获取自 UnixEpoch 开始指定时间间隔 d 后的时间。

- public static func of(year!: Int64, month!: Int64, dayOfMonth!: Int64, hour!: Int64 = 0, minute!: Int64 = 0, second!: Int64 = 0, nanosecond!: Int64 = 0, timeZone!: TimeZone=TimeZone.Local): DateTime

根据参数给定的年、月、日、时、分、秒、纳秒、时区构造 DateTime 实例,所有参数均不能超过其正常范围,否则会抛出异常,参数的正常范围如下所示。

(1) year:年,范围为[-999,999,999, 999,999,999]。

(2) month:月,范围为[1, 12]。

(3) dayOfMonth:日,范围为[1, 31],最大取值需要跟月份匹配,可能是 28、29、30、31。

(4) hour:时,范围为[0, 23]。

(5) minute:分,范围为[0, 59]。

(6) second:秒,范围为[0, 59]。

(7) nanosecond:纳秒,范围为[0, 999,999,999]。

- public static func of(year!: Int64, month!: Month, dayOfMonth!: Int64, hour!: Int64 = 0, minute!: Int64 = 0, second!: Int64 = 0, nanosecond!: Int64 = 0, timeZone!: TimeZone=TimeZone.Local): DateTime

根据参数给定的年、月、日、时、分、秒、纳秒、时区构造 DateTime 实例,该函数和上一个函数类似,只是月份的参数类型变成了 Month。

- public static func ofUTC(year!: Int64, month!: Int64, dayOfMonth!: Int64, hour!: Int64 = 0, minute!: Int64 = 0, second!: Int64 = 0, nanosecond!: Int64 = 0): DateTime

根据参数给定的年、月、日、时、分、秒、纳秒构造 UTC 时区 DateTime 实例,所有参数均不能超过其正常范围,否则会抛出异常。

- public static func ofUTC(year!: Int64, month!: Month, dayOfMonth!: Int64, hour!: Int64 = 0, minute!: Int64 = 0, second!: Int64 = 0, nanosecond!: Int64 = 0): DateTime

根据参数给定的年、月、日、时、分、秒、纳秒构造 UTC 时区 DateTime 实例,该函数和上一个函数类似,只是月份的参数类型变成了 Month。

- public static func parse(str: String): DateTime

从参数 str 解析得到 DateTime 实例,str 的格式默认为 RFC3339 规定的格式,无法正常解析时,抛出异常。

- public static func parse(str: String, format: String): DateTime

从参数 str 解析得到 DateTime 实例,str 的格式为 format 指定的格式,无法正常解析时,抛出异常。

构造随机时间实例的示例代码如下:

```
//Chapter19/part_of_time_demo.cj

from std import time.*
from std import random.*

main() {
    let rand = Random()

    //使用随机值生成时间
    let curTime = DateTime.of(year: 2000 + rand.nextInt64(20), month: 1 + rand.nextInt64(12),
        dayOfMonth: 1 + rand.nextInt64(28), hour: rand.nextInt64(24), minute: rand.nextInt64(60),
        second: rand.nextInt64(60), nanosecond: rand.nextInt64(1000), timeZone: TimeZone.Local)

    //按照年-月-日时:分:秒的格式打印时间
    println(
        "${curTime.year} - ${curTime.monthValue} - ${curTime.dayOfMonth} ${curTime.hour}:${curTime.minute}:${curTime.second}"
    )
}
```

编译后运行该示例,可能的输出如下:

```
2019 - 3 - 29 3:16:36
```

3．增加时间的相关函数
- public func addYears(n：Int64)：DateTime

获取参数 n 年之后的时间，返回新的实例，新的实例为合法的日期，例如，对于 2024-02-29 这样的时间，增加一年后会变为 2025-02-28。

- public func addMonths(n：Int64)：DateTime

获取参数 n 月之后的时间，返回新的实例，新的实例为合法的日期。

- public func addWeeks(n：Int64)：DateTime

获取参数 n 周之后的时间，返回新的实例。

- public func addDays(n：Int64)：DateTime

获取参数 n 天之后的时间，返回新的实例。

- public func addHours(n：Int64)：DateTime

获取参数 n 小时之后的时间，返回新的实例。

- public func addMinutes(n：Int64)：DateTime

获取参数 n 分钟之后的时间，返回新的实例。

- public func addSeconds(n：Int64)：DateTime

获取参数 n 秒之后的时间，返回新的实例。

- public func addNanoseconds(n：Int64)：DateTime

获取参数 n 纳秒之后的时间，返回新的实例。

4．重载操作符函数
- public operator func ＋(r：Duration)：DateTime

返回一个新实例，表示参数 r 指定时间间隔后的时间。

- public operator func －(r：Duration)：DateTime

返回一个新实例，表示参数 r 指定时间间隔前的时间。

- public operator func －(r：DateTime)：Duration

返回一个 Duration 实例，表示两个 DateTime 实例相距的时间间隔，如果参数 r 表示的时间早于本实例，结果为正，晚于为负，等于为 0。

- public operator func ＝＝(r：DateTime)：Bool

当前实例是否等于参数 r。

- public operator func ！＝(r：DateTime)：Bool

当前实例是否不等于参数 r。

- public operator func ＞＝(r：DateTime)：Bool

当前实例是否晚于或等于参数 r。

- public operator func ＞(r：DateTime)：Bool

当前实例是否晚于参数 r。

- public operator func ＜＝(r：DateTime)：Bool

当前实例是否早于或等于参数 r。

- public operator func <(r: DateTime): Bool

当前实例是否早于参数 r。

操作符重载的示例代码如下:

```
//Chapter19/operator_time_demo.cj

from std import time.*
from std import random.*

main() {
    let rand = Random()

    //获取当前时间
    let locTime = DateTime.now()

    //在当前时间加上随机的年月日
    let newTime1 = locTime.addYears(rand.nextInt64(3)).addMonths(rand.nextInt64(10)).addDays(rand.nextInt64(10))

    //生成一个随机的时间间隔
    let duration = Duration.hour * rand.nextInt64(10000) + Duration.second * rand.nextInt64(10000)

    //在当前时间上加上一个随机的时间间隔
    let newTime2 = locTime + duration

    //计算两个时间的差值
    let newDuration = newTime2 - newTime1

    //输出时间差值
    println(newDuration)
}
```

编译后运行该示例,可能的输出如下:

```
119d22h32m36s
```

### 5. 其他函数

- public static func now(timeZone!: TimeZone=TimeZone.Local): DateTime

获取参数 timeZone 指定时区的当前时间,默认为当前时区。

- public static func nowUTC(): DateTime

获取 UTC 时区的当前时间。

- public func inUTC(): DateTime

当前实例在 UTC 时区的时间。

- public func inLocal(): DateTime

当前实例在本地时区的时间。

- public func inTimeZone(timeZone：TimeZone)：DateTime

当前实例在参数 timeZone 指定时区的时间。

- public func compare(rhs：DateTime)：Ordering

判断一个当前实例与参数 rhs 的大小关系，如果大于，则返回 Ordering.GT；如果等于，则返回 Ordering.EQ；如果小于，则返回 Ordering.LT。

- public func toString()：String

当前实例的字符串形式，格式为 RFC3339 规定的格式。

- public func toString(format：String)：String

当前实例的字符串形式，格式由参数 format 指定。

## 19.8.6　时间格式

在 DateTime 实例转换为字符串，或者字符串解析为 DateTime 实例时都可以指定时间字符串的格式，例如 yyyy-MM-dd HH:mm:ss OO 等，在这个格式字符串里，不同的字符具有不同的含义，下面按照时间的不同部分进行解析。

### 1. 年

使用 1～9 位的字符 Y 表示，Y 可以是大写也可以是小写；不同的位数表示如下：

- 1、3、4 位时，表示 4 位的年份，如 2024；
- 2 位时，表示 2 位的年份，如 24；
- 其他位数时，按照实际位数解析，位数不足时前面补 0。

年的格式化示例代码如下：

```
//Chapter19/year_format_demo.cj

from std import time.*

main() {
    //构造 2024 年 1 月 1 日 8:30 的时间
    let time = DateTime.of(year: 2024, month: 1, dayOfMonth: 1, hour: 8, minute: 30)

    //输出格式字符串为 1～9 位 y 时的时间表示形式
    for (i in 1..=9) {
        let fmt = String(Array<Char>(i, item: 'y'))
        println("${fmt}: ${time.toString(fmt)}")
    }
}
```

编译后运行该示例，输出如下：

```
y: 2024
yy: 24
yyy: 2024
yyyy: 2024
```

```
yyyyy: 02024
yyyyyy: 002024
yyyyyyy: 0002024
yyyyyyyy: 00002024
yyyyyyyyy: 000002024
```

### 2. 月

使用 1~4 位大写的字符 M 表示，不同的位数表示如下：

- 1 位时，表示 1~2 位的月份，如 1、02、11 等；
- 2 位时，表示 2 位的月份，如 01、11 等；
- 3 位时，表示月份的简写，如 Jan、Dec 等；
- 4 位时，表示月份的完整书写，如 January、December 等。

月的格式化示例代码如下：

```
//Chapter19/year_format_demo.cj

from std import time.*

main() {
    //构造 2024 年 1 月 1 日 8:30 的时间
    let time = DateTime.of(year: 2024, month: 1, dayOfMonth: 1, hour: 8, minute: 30)

    //输出格式字符串为 1~4 位 M 时的时间表示形式
    for (i in 1..=4) {
        let fmt = String(Array<Char>(i, item: 'M'))
        println(" ${fmt}: ${time.toString(fmt)}")
    }
}
```

编译后运行该示例，输出如下：

```
M: 1
MM: 01
MMM: Jan
MMMM: January
```

### 3. 日

使用 1~2 位小写的字符 d 表示，不同的位数表示如下：

- 1 位时，表示 1 到 2 位的日期，如 1、02、11 等；
- 2 位时，表示 2 位的日期，如 01、11 等。

### 4. 时

使用 1~2 位大写的字符 H 表示 24 小时制；使用 1 到 2 位小写的字符 h 表示 12 小时制，不同的位数表示如下：

- 1 位 H 时，表示 1 到 2 位的小时，如 0、00、23 等；
- 1 位 h 时，表示 1 到 2 位的小时，如 1、01、12 等；

- 2 位 H 时，表示 2 位的小时，如 00、23 等；
- 2 位 h 时，表示 2 位的小时，如 01、12 等。

时的格式化示例代码如下：

```
//Chapter19/hour_format_demo.cj

from std import time.*

main() {
    //构造 2024 年 1 月 1 日 0:30 的时间
    var time = DateTime.of(year: 2024, month: 1, dayOfMonth: 1, hour: 0, minute: 30)
    printFormat(time)

    //构造 2024 年 1 月 1 日 11:30 的时间
    time = DateTime.of(year: 2024, month: 1, dayOfMonth: 1, hour: 11, minute: 30)
    printFormat(time)

    //构造 2024 年 1 月 1 日 23:30 的时间
    time = DateTime.of(year: 2024, month: 1, dayOfMonth: 1, hour: 23, minute: 30)
    printFormat(time)
}

func printFormat(time: DateTime) {
    //输出格式字符串为 1~2 位 H 时的时间表示形式
    for (i in 1..=2) {
        let fmt = String(Array<Char>(i, item: 'H'))
        println("${fmt}: ${time.toString(fmt)}")
    }

    //输出格式字符串为 1 到 2 位 h 时的时间表示形式
    for (i in 1..=2) {
        let fmt = String(Array<Char>(i, item: 'h'))
        println("${fmt}: ${time.toString(fmt)}")
    }
}
```

编译后运行该示例，输出如下：

```
H: 0
HH: 00
h: 12
hh: 12
H: 11
HH: 11
h: 11
hh: 11
H: 23
HH: 23
h: 11
hh: 11
```

**5. 分**

使用 1～2 位小写的字符 m 表示，不同的位数表示如下：
- 1 位时，表示 1～2 位的分钟，如 0、1、02、59 等；
- 2 位时，表示 2 位的分钟，如 00、01、02、59 等。

**6. 秒**

使用 1～2 位小写的字符 s 表示，不同的位数表示如下：
- 1 位时，表示 1～2 位的秒数，如 0、1、02、59 等；
- 2 位时，表示 2 位的秒数，如 00、01、02、59 等。

**7. 纳秒**

使用 1～3 位大写的字符 S 表示，不同的位数表示如下：
- 1 位时，表示 3 位的毫秒，如 001、123 等；
- 2 位时，表示 6 位的微秒，如 000001、123456 等；
- 3 位时，表示 9 位的纳秒，如 000000001、123456789 等。

**8. 时区**

表示时区的字符有大写的 O、小写的 z 和大写的 Z，分别表示时区的不同形式，依次为偏移形式、时区名和地点形式及 GMT 偏移形式都可以使用 1 到 4 位的字符，下面分别进行说明具体用法。

(1) 大写的 O，不同的位数表示如下：
- 1 位时，输出偏移量的小时部分，如+08、−08 等，若偏移量不为整小时则会截断；
- 2 位时，输出偏移量的小时及分钟部分，如+08:00、−08:59 等，若偏移量不为整分钟则会截断；
- 3 位时，输出偏移量的时分秒部分，如+08:00:00、−08:59:59 等；
- 4 位时，如果偏移量包括秒，则参照 3 位时输出；如果不包括秒，则参照 2 位时输出；如果偏移量为 0，则输出 Z。

(2) 小写的 z：

1～4 位时都是显示时区名+偏移量，如果偏移量为 0，则不显示加号和偏移量；如果偏移量的秒为 0，则不显示秒部分；如果偏移量的分钟和秒都为 0，则不显示分钟和秒部分；在显示偏移量时，也不包括时、分、秒中间的冒号。例如，对于偏移量+08:30:01，显示的内容为 GMT+083001。

(3) 大写的 Z，不同的位数表示如下：
- 1 位时，输出 GMT+偏移量的小时部分，如+08、−08 等，若偏移量不为整小时则会截断；
- 2 位时，输出 GMT+偏移量的小时及分钟部分，如+08:00、−08:59 等，若偏移量不为整分钟则会截断；
- 3 位时，输出 GMT+偏移量的时分秒部分，如+08:00:00、−08:59:59 等；
- 4 位时，如果偏移量包括秒参照 3 位时输出，如果不包括秒，则参照 2 位时输出。

时区的格式化示例代码如下：

```
//Chapter19/timezone_format_demo.cj

from std import time.*

main() {
    //构造东八区2024年1月1日8:30的时间
    let time = DateTime.parse("2024-01-01T08:30:00+08:30:00")

    //输出格式字符串为1~4位M时的时间表示形式
    for (i in 1..=4) {
        let fmt = String(Array<Char>(i, item: 'O'))
        println("${fmt}: ${time.toString(fmt)}")
    }

    for (i in 1..=4) {
        let fmt = String(Array<Char>(i, item: 'z'))
        println("${fmt}: ${time.toString(fmt)}")
    }

    for (i in 1..=4) {
        let fmt = String(Array<Char>(i, item: 'Z'))
        println("${fmt}: ${time.toString(fmt)}")
    }
}
```

编译后运行该示例，输出如下：

```
O: +08
OO: +08:30
OOO: +08:30:00
OOOO: +08:30
z: GMT+0830
zz: GMT+0830
zzz: GMT+0830
zzzz: GMT+0830
Z: GMT+08
ZZ: GMT+08:30
ZZZ: GMT+08:30:00
ZZZZ: GMT+08:30
```

### 9. 一年中的第几天

使用1~2位大写的字符 D 表示，不同的位数表示如下：
- 1位时，输出1~3位数字，例如1、12、123等；
- 2位时，输出3位数字，例如001、012、123等。

### 10. 周

使用1~2位大写的字符 W 表示，不同的位数表示如下：

- 1位时,输出 1 到 2 位数字,例如 1、12 等;
- 2位时,输出 2 位数字,例如 01、12 等。

### 11. 一周中的第几天

使用 1~4 位小写的字符 w 表示,不同的位数表示如下:

- 1位时,输出 1 位数字,0 表示周日,1~6 分别表示周一到周六;
- 2位时,输出 2 位数字,00 表示周日,01~06 分别表示周一到周六;
- 3位时,输出简写的周,如 Sun、Sat 等;
- 4位时,输出完整拼写的周,如 Sunday、Saturday 等。

### 12. 纪年

使用 1~3 位大写的字符 G 表示,不同的位数表示如下:

- 1位时,输出 A;
- 2位时,输出 AD;
- 3位时,输出 Anno Domini。

### 13. 上下午

使用小写字符 a 表示,输出 AM 或 PM。时间的格式化解析和格式化输出示例代码如下:

```
//Chapter19/parse_demo.cj

from std import time.*

main() {
    //符合 RFC3339 格式的时间字符串
    let rfcFmtStr = "2023-05-01T20:00:00+08:00"

    //自定义时间格式和时间字符串
    let custFmt = "yyyy-MM-dd HH:mm:ssOOOO"
    let custFmtStr = "2024-05-01 20:30:00+08:00"

    //解析时间字符串
    let rfcTime = DateTime.parse(rfcFmtStr)
    let custTime = DateTime.parse(custFmtStr, custFmt)

    let outFmt = "GG yyyy年 MM月 dd日 a hh时 mm分 ss秒 wwww"

    //按照自定义格式输出
    println(rfcTime.toString(outFmt))
    println(custTime.toString(outFmt))
}
```

编译后运行该示例,输出如下:

```
AD 2023年 05月 01日 PM 08时 00分 00s Monday
AD 2024年 05月 01日 PM 08时 30分 00s Wednesday
```

# 第 20 章 字符及字符串处理

字符(Char)及字符串(String)的处理是软件开发中基础的功能,本节将介绍这两种类型的常用处理函数,通过示例代码的方式帮助读者快速掌握常用用法。

## 20.1 字符处理

仓颉库提供了 Char 类型的扩展函数,主要扩展函数如下。
- public func isAsciiLetter():Bool

判断字符是否为 ASCII 码字母。
- public func isAsciiNumber():Bool

判断字符是否为 ASCII 码数字。
- public func isAsciiLowerCase():Bool

判断字符是否为小写字母。
- public func isAsciiUpperCase():Bool

判断字符是否为大写字母。
- public func isAsciiWhiteSpace():Bool

判断字符是否为空白字符。
- public func toAsciiUpperCase():Char

将字符转换为 ASCII 码大写字母。
- public func toAsciiLowerCase():Char

将字符转换为 ASCII 码小写字母。

函数使用的示例代码如下:

```
//Chapter20/char_demo.cj

from std import console.*

main() {
    println("Please enter a character,'q' to exit:")
```

```
        //读取输入字符,如果输入出错,则置成字符q
        var chr = Console.stdIn.read() ?? 'q'

        //输出的字符为q就退出
        while (chr != 'q') {
            //如果字符是空白字符,则不处理,然后接收新的输入并继续
            if (!chr.isAsciiWhiteSpace()) {
                //如果是数字,提示输入的数字
                if (chr.isAsciiNumber()) {
                    println("You entered a number ${chr}")
                } else {
                    //如果是小写字母,转换为大写
                    if (chr.isAsciiLowerCase()) {
                        chr = chr.toAsciiUpperCase()
                    }

                    //输出字符
                    println("The character is ${chr}")
                }
            }

            //接收新的输入字符
            chr = Console.stdIn.read() ?? 'q'
        }
    }
```

编译后运行该示例,依次输入1、a、+、q,输出如下:

```
Please enter a character,q to exit:
1
You entered a number 1
a
The character is A
+
The character is +
q
```

## 20.2 字符串处理

### 20.2.1 字符串转数组

- public func toUtf8Array(): Array<UInt8>
- public func toUtf16Array(): Array<UInt16>
- public func toUtf32Array(): Array<UInt32>
- public func toCharArray(): Array<Char>

返回字符串实例的 Utf8、Utf16、Utf32 表示的数组及字符数组。

## 20.2.2 指定位置字符获取

- public func get(index：Int64)：Option＜Char＞

获取下标 index 对应的字符值，如果 index 小于 0、大于或等于字符串长度，则返回 Option＜Char＞.None。该函数获取返回值的效率较低，不建议在循环中使用。

## 20.2.3 子字符串获取

- public func substring(beginIndex：Int64)：String

获取字符串从索引 beginIndex 开始到结束的子字符串，beginIndex 必须大于或等于 0，并且小于原字符串长度，否则会抛出异常。

- public func substring(beginIndex：Int64, length：Int64)：String

获取字符串从索引 beginIndex 开始长度为 length 的子字符串，beginIndex 必须大于或等于 0，并且小于原字符串长度，否则会抛出异常；length 小于 0 或者（beginIndex + length）的值大于原字符串的长度也会抛出异常。

示例代码如下：

```
//Chapter20/substring_demo.cj

main() {
    let oriString = "Cangjie is the best programming language!"

    //获取从索引 11 开始的子字符串
    let rightStr = oriString.substring(11)

    //获取从索引 15 开始长度为 4 的子字符串
    let midStr = oriString.substring(15, 4)

    //打印字符串
    println(rightStr)
    println(midStr)
}
```

编译后运行该示例，输出如下：

```
the best programming language!
best
```

## 20.2.4 字符查找

- public func indexOf(c：Char)：Option＜Int64＞

查找参数 c 代表的字符在字符串内第 1 次出现的索引，如果字符串中没有此字符，则返回 Option＜Int64＞.None。

- public func indexOf(c：Char，fromIndex：Int64)：Option＜Int64＞

从字符串的 fromIndex 下标处开始查找参数 c 代表的字符，返回第 1 次出现的索引；如果 fromIndex 大于或等于字符串的长度或者在 fromIndex 下标及其之后没有出现此字符，则返回 Option＜Int64＞.None。

- public func lastIndexOf(c：Char)：Option＜Int64＞

查找参数 c 代表的字符在字符串内最后一次出现的索引，如果字符串中没有此字符，则返回 Option＜Int64＞.None。

- public func lastIndexOf(c：Char，fromIndex：Int64)：Option＜Int64＞

从字符串的 fromIndex 下标处开始查找参数 c 代表的字符，返回最后一次出现的索引，如果 fromIndex 大于或等于字符串的长度或者在 fromIndex 下标及其之后没有出现此字符，则返回 Option＜Int64＞.None。

- public func count(c：char)：Int64

返回参数 c 代表的字符在原字符串中出现的次数。

查找字符位置的示例代码如下：

```
//Chapter20/find_char_demo.cj

main() {
    //定义原始字符串
    let oriString = "Cangjie is the best programming language!"

    //查找字符 A 的位置
    let posA = oriString.indexOf('A')

    //打印字符位置
    match (posA) {
        case Some(pos) => println("The first occurrence of character A is ${pos}!")
        case None => println("The character A does not appear!")
    }

    //查找字符 C 的位置
    let posC = oriString.indexOf('C')

    //打印字符位置
    match (posC) {
        case Some(pos) => println("The first occurrence of character C is ${pos}!")
        case None => println("The character C does not appear!")
    }

    //从索引 11 开始查找字符 a 最后出现的位置
    let lastPosa = oriString.lastIndexOf('a', 11)

    //打印字符位置
    match (lastPosa) {
```

```
            case Some(pos) => println("The last occurrence of character a is ${pos}!")
            case None => println("The character a does not appear!")
    }
}
```

编译后运行该示例,输出如下:

```
The character A does not appear!
The first occurrence of character C is 0!
The last occurrence of character a is 37!
```

## 20.2.5 子字符串查找

- public func indexOf(sub: String): Option<Int64>

查找子字符串 sub 在字符串中第 1 次出现的起始索引,如果原字符串中没有 sub 字符串,则返回 Option<Int64>.None。

- public func indexOf(sub: String, fromIndex: Int64): Option<Int64>

从字符串的 fromIndex 下标处开始查找子字符串 sub,返回第 1 次出现的起始索引;如果 fromIndex 大于或等于原字符串的长度或者在 fromIndex 下标及其之后没有出现子字符串 sub,则返回 Option<Int64>.None。

- public func lastIndexOf(sub: String): Option<Int64>

查找子字符串 sub 在字符串中最后一次出现的起始索引,如果原字符串中没有 sub 字符串,则返回 Option<Int64>.None。

- public func lastIndexOf(sub: String, fromIndex: Int64): Option<Int64>

从字符串的 fromIndex 下标处开始查找子字符串 sub,返回最后一次出现的起始索引;如果 fromIndex 大于或等于原字符串的长度或者在 fromIndex 下标及其之后没有出现子字符串 sub,则返回 Option<Int64>.None。

查找子字符串位置的示例代码如下:

```
//Chapter20/find_substring_demo.cj

main() {
    //定义原始字符串
    let oriString = "Cangjie is the best programming language!"

    //查找字符串 best 的位置
    let posBest = oriString.indexOf("best")

    //打印字符串位置
    match (posBest) {
        case Some(pos) => println("The first occurrence of the string \'best\' is ${pos}!")
        case None => println("The string \'best\' does not appear!")
    }
```

```
        //从索引 11 开始查找字符串 the 最后出现的位置
        let lastPos = oriString.lastIndexOf("the", 11)

        //打印字符位置
        match (lastPos) {
            case Some(pos) => println("The last occurrence of the string \'the\' is ${pos}!")
            case None => println("The string \'the\'\' does not appear!")
        }
}
```

编译后运行该示例,输出如下:

```
The first occurrence of the string 'best' is 15!
The last occurrence of the string 'the' is 11!
```

### 20.2.6 字符串修整

- public func trimAscii(): String

返回原字符串去掉了首尾空白字符后的新字符串。

- public func trimAsciiLeft(): String

返回原字符串去掉了开头的空白字符后的新字符串。

- public func trimAsciiRight(): String

返回原字符串去掉了结尾的空白字符后的新字符串。

- public func trimLeft(prefix: String): String

返回原字符串去掉前缀 prefix 后的新字符串。

- public func trimRight(suffix: String): String

返回原字符串去掉后缀 suffix 后的新字符串。

去除首尾特定字符的示例代码如下:

```
//Chapter20/trim_demo.cj

main() {
    let oriString = " Cangjie is the best programming language! "

    //去除首尾的空白字符
    let trimStr = oriString.trimAscii()

    //去除开头的空白字符
    let trimStart = oriString.trimAsciiLeft()

    //先去除字符串仓颉,然后去除开头的空白字符
    let trimPrefix = trimStr.trimLeft("Cangjie").trimAsciiLeft()

    //打印字符串
    println(trimStr)
```

```
    //为显示最后的空格加上字符串"end"
    println(trimStart + "end")
    println(trimPrefix)
}
```

编译后运行该示例,输出如下:

```
Cangjie is the best programming language!
Cangjie is the best programming language! end
is the best programming language!
```

## 20.2.7　字符串分隔

- public func split(c: Char, removeEmpty!: Bool=false): Array<String>

对原字符串按照参数 c 代表的字符分隔符进行分割,返回分割后的子字符串数组,当 c 未出现在原字符串中时,返回长度为 1 的字符串数组,唯一的元素为原字符串;removeEmpty 表示是否移除分割结果中的空字符串,默认值为 false。

- public func split(c: Char, maxSplit: Int64, removeEmpty!: Bool=false): Array<String>

对原字符串按照参数 c 代表的字符分隔符进行分割,最多分割为 maxSplit 个子字符串,并返回分割后的子字符串数组;当 c 未出现在原字符串中时,返回长度为 1 的字符串数组,唯一的元素为原字符串;当 maxSplit 大于完整分割出来的子字符串数量时,返回完整分割的字符串数组;当对原字符串按照字符分隔符 c 进行分割的子字符串数量大于 maxSplit 时,前 maxSplit-1 个子字符串正常分割,剩余的所有字符串作为最后 1 个子字符串;removeEmpty 表示是否移除分割结果中的空字符串,默认值为 false。

- public func split(str: String, removeEmpty!: Bool=false): Array<String>

对原字符串按照字符串 str 进行分割,返回分割后的子字符串数组,当 str 未出现在原字符串中时,返回长度为 1 的字符串数组,唯一的元素为原字符串;removeEmpty 表示是否移除分割结果中的空字符串,默认值为 false。

- public func split(str: String, maxSplit: Int64, removeEmpty!: Bool=false): Array<String>

对原字符串按照字符串 str 进行分割,最多分割为 maxSplit 个子字符串,并返回分割后的子字符串数组;当 str 未出现在原字符串中时,返回长度为 1 的字符串数组,唯一的元素为原字符串;当 maxSplit 大于完整分割出来的子字符串数量时,返回完整分割的字符串数组;当对原字符串按照字符串 str 进行分割的子字符串数量大于 maxSplit 时,前 maxSplit-1 个子字符串正常分割,剩余的所有字符串作为最后 1 个子字符串;removeEmpty 表示是否移除分割结果中的空字符串,默认值为 false。

字符串分隔的示例代码如下:

```
//Chapter20/split_demo.cj
```

```
main() {
    let oriString =
    """
    北国风光,千里冰封,万里雪飘.
    望长城内外,惟余莽莽;大河上下,顿失滔滔.
    山舞银蛇,原驰蜡象,欲与天公试比高.
    须晴日,看红装素裹,分外妖娆.
    江山如此多娇,引无数英雄竞折腰.
    惜秦皇汉武,略输文采;唐宗宋祖,稍逊风骚.
    一代天骄,成吉思汗,只识弯弓射大雕.
    俱往矣,数风流人物,还看今朝."""

    //使用句号分隔字符串
    var sentenceArray = oriString.split('.')

    var sn = 1
    for (sentenc in sentenceArray) {
        //打印分隔后去掉前后空白字符的子字符串,并在前面加上序号
        println(" ${sn}:${sentenc.trimAscii()}")
        sn++
    }

    //使用句号分隔字符串,最多分隔4个子字符串
    sentenceArray = oriString.split('.', 4)
    sn = 1
    for (sentenc in sentenceArray) {
        println(" ${sn}:${sentenc.trimAscii()}")
        sn++
    }

    let newStr =
    "一个人能力有大小,但只要有这点精神,就是一个高尚的人,一个纯粹的人,一个有道德的人,一个脱离了低级趣味的人,一个有益于人民的人."

    let splitArray = newStr.split("一个", 4)

    sn = 1
    for (subStr in splitArray) {
        println(" ${sn}:${subStr}")
        sn++
    }
}
```

编译后运行该示例,输出如下:

```
1:北国风光,千里冰封,万里雪飘
2:望长城内外,惟余莽莽;大河上下,顿失滔滔
3:山舞银蛇,原驰蜡象,欲与天公试比高
4:须晴日,看红装素裹,分外妖娆
5:江山如此多娇,引无数英雄竞折腰
```

```
6:惜秦皇汉武,略输文采;唐宗宋祖,稍逊风骚
7:一代天骄,成吉思汗,只识弯弓射大雕
8:俱往矣,数风流人物,还看今朝
9:
1:北国风光,千里冰封,万里雪飘
2:望长城内外,惟余莽莽;大河上下,顿失滔滔
3:山舞银蛇,原驰蜡象,欲与天公试比高
4:须晴日,看红装素裹,分外妖娆.
    江山如此多娇,引无数英雄竞折腰.
    惜秦皇汉武,略输文采;唐宗宋祖,稍逊风骚.
    一代天骄,成吉思汗,只识弯弓射大雕.
    俱往矣,数风流人物,还看今朝.
1:
2:人能力有大小,但只要有这点精神,就是
3:高尚的人,
4:纯粹的人,一个有道德的人,一个脱离了低级趣味的人,一个有益于人民的人.
```

从上述示例可以看出,如果分隔符或者进行分隔的字符串在原字符串起始的位置,则在返回的子字符串数组中,第 1 个元素是空字符串;如果在原字符串结尾的位置,则在返回的子字符串数组里,最后一个元素是空字符串.

### 20.2.8 字符串判断

- public func contains(c: Char): Bool

判断原字符串中是否包含参数 c 代表的字符。

- public func contains(str: String): Bool

判断原字符串中是否包含字符串 str,如果 str 字符串长度为 0,则返回值为 true。

- public func startsWith(prefix: String): Bool

判断原字符串是否以 prefix 字符串为前缀开始,当 prefix 长度为 0 时,返回值为 true。

- public func endsWith(suffix: String): Bool

判断原字符串是否以 suffix 字符串为后缀结尾,当 suffix 长度为 0 时,返回值为 true。

- public func equals(str: String): Bool

比较原字符串是否和字符串 str 的值相同。

- public func isEmpty(): Bool

判断原字符串是否为空字符串。

- public func isAsciiBlank(): Bool

当原字符串为空或者包含的字符都为 ASCII 码中的空白字符(包括 0x09、0x10、0x11、0x12、0x13、0x20)时,返回值为 true,否则返回值为 false。

### 20.2.9 字符串连接

- public func concat(str: String): String

将指定字符串 str 连接到原字符串的末尾并返回连接后的新字符串。

- public operator func +(right: Char): String

重载"+"操作符,把 right 字符拼接到原字符串的末尾并返回拼接后的新字符串。

- public operator func +(right: String): String

重载"+"操作符,把 right 字符串拼接到原字符串的末尾并返回拼接后的新字符串。

- public operator func *(count: Int64): String

重载"*"操作符,原字符串重复 count 次并作为新字符串返回。

- public func padLeft(totalWidth: Int64, paddingChar!: Char = ' '): String

按指定长度右对齐原字符串,不足的部分在左侧使用指定的字符 paddingChar 填充。如果 totalWidth 小于原字符串长度,则返回原字符串,不会截断。

- public func padRight(totalWidth: Int64, paddingChar!: Char = ' '): String

按指定长度左对齐原字符串,不足的部分在右侧使用指定的字符 paddingChar 填充。如果 totalWidth 小于原字符串长度,则返回原字符串,不会截断。

- public static func join(strArray: Array<String>, delimiter!: String = String.empty): String

使用中间字符串 delimiter 连接字符串数组 strArray 中的所有字符串,并作为新字符串返回。

字符串连接的示例代码如下:

```
//Chapter20/join_demo.cj

from std import time.*
from std import console.*

main() {
    //定义可以容纳 5 个字符串的字符串数组
    let strArray = Array<String>(5, item: "")
    println("Please enter 5 strings:")
    var index = 0

    //让用户输入 5 个字符串,并存储到字符串数组 strArray 中,如果输入有问题,则置为 demo 字
    //符串
    while (index < 5) {
        let input = Console.stdIn.readln() ?? "demo"
        strArray[index] = input
        index++
    }

    //定义可变的字符串变量
    var newStr = ""

    //循环次数
    let circleCount = 1000000

    //循环计数开始
```

```
    index = 0

    //记录concat函数连接开始时间
    var startTime = DateTime.now()

    //循环一百万次,也就是把字符串数组strArray中的字符串连接成一个字符串100万次
    while (index < circleCount) {
        newStr = ""

        //使用concat方法连接字符串
        for (str in strArray) {
            newStr = newStr.concat(str)
        }
        index++
    }

    //记录结束时间,计算使用concat函数连接的用时
    var costTime1 = DateTime.now() - startTime

    index = 0

    //记录join函数连接开始时间
    startTime = DateTime.now()

    //循环一百万次,也就是把字符串数组strArray中的字符串连接成一个字符串100万次
    while (index < circleCount) {
        newStr = String.join(strArray)
        index++
    }

    //记录结束时间,计算使用join函数连接的用时
    var costTime2 = DateTime.now() - startTime

    //输出两次测试的用时和时间差
    println("concat function: ${costTime1}; join function: ${costTime2};The difference is ${costTime1 - costTime2}")
}
```

编译后运行该示例,并按照提示输入5个字符串,假如输入的字符串依次为cangjie、is、the best、programming、language,可能的输出如下:

```
Please enter 5 strings:
cangjie
is
the best
programming
language
concat function:654ms797us300ns;join function:125ms751us100ns;The difference is 529ms46us200ns
```

从上述示例可以看出,使用join()连接字符串的效率远高于使用concat(),这是因为字

符串本身是不可改变的,concat()函数每连接一个字符串都要创建一个新的字符串,而join()可以把字符串数组中的所有字符串一次性地连接成一个新字符串,所以后者效率更高一些。

### 20.2.10 字符串替换与反转

- public func replace(oldChar: Char, newChar: Char): String

使用字符 newChar 替换原字符串中的字符 oldChar 并返回替换后的新字符串。

- public func replace(oldStr: String, newStr: String): String

使用字符串 newStr 替换原字符串中的字符串 oldStr 并返回替换后的新字符串。

- public func reverse(): String

返回字符串反转后的新字符串。

## 20.3 猜数字小游戏

本节将通过一个稍微复杂一点的小游戏,综合应用前几个章节学习过的知识,重点演示字符及字符串的处理。这个小游戏由系统产生一个大于或等于 0 但小于 100 的随机整数,游戏玩家可以向系统提出问题,系统只会回答 Yes 或者 No,最终游戏玩家根据系统的提示猜测那个随机数。游戏玩家提问题的方式是限定的,只能使用"判断符号+数字"的形式,例如">=50",表示提问系统那个数字是否大于或等于 50;"<30"表示那个数字是否小于 30;"20""=20""==20"都表示那个数字是 20,为了简化起见,判断时只能使用>、>=、==、<、<=与数字的组合,示例代码如下:

```
//Chapter20/figure_guessing_game.cj

from std import console.*
from std import random.Random
from std import convert.*

//游戏运行标志,默认为不运行
var gameRunning: Bool = false

//要猜测的数字
var guessedNumber = 0

//问问题的次数
var askTimes = 0

//操作符号和数字类
class OpAndNum {
    //op:操作符号    num:数字
    OpAndNum(let op: String, let num: Int64) {}
}

//从输入的字符串中分析出操作符号和数字
```

```
func getOpAndNum(input: String): Option<OpAndNum> {
    //如果输入字符串为空,则直接返回 None
    if (input.isEmpty()) {
        return Option<OpAndNum>.None
    }

    //如果第 1 位是数字,则表明整个 input 都是数字,并且是判断相等的操作
    if (input[0].isAsciiNumber()) {
        return createOpAndNum("==", input)
    }

    //因为上面判断了第 1 位是不是数字,如果不是数字并且长度只有 1 位,则肯定是非法输入
    if (input.size == 1) {
        return Option<OpAndNum>.None
    }

    //取前两位作为符号字符串
    let sign2Char = input[0..=1]

    //判断符号字符串是不是给定的符号组合
    if (sign2Char.equals("==") || sign2Char.equals(">=") || sign2Char.equals("<=")) {
        //input 长度必须大于 2,否则没有数字位
        if (input.size > 2) {
            return createOpAndNum(sign2Char, input[2..])
        } else {
            return Option<OpAndNum>.None
        }
    }

    //判断第 1 位是不是给定的符号
    if (input[0] == '=' || input[0] == '>' || input[0] == '<') {
        //input 长度必须大于 1,否则没有数字位
        if (input.size > 1) {
            return createOpAndNum(input[0].toString(), input[1..])
        } else {
            return Option<OpAndNum>.None
        }
    }

    //如果不符合上述几种情况,则返回 None
    return Option<OpAndNum>.None
}

//解析表示数字的字符串参数 numStr 得到整数,结合给定的符号参数 sign,生成
//Option<OpAndNum>对象
func createOpAndNum(sign: String, numStr: String): Option<OpAndNum> {
    //把 numStr 去掉首尾的空白字符后转换成整数
    let numOption = Int64.tryParse(numStr.trimAscii())
    match (numOption) {
```

```
            //如果转换成功,则返回 Some 枚举
            case Some(value) => return Option<OpAndNum>.Some(OpAndNum(sign, value))

            //如果转换失败,则返回 None 枚举
            case None => return Option<OpAndNum>.None
        }
    }

//开始一局新游戏
func startNewGame() {
    //产生一个随机数,作为本局游戏要猜测的数字
    guessedNumber = Random().nextInt64(100)

    //设置游戏运行标志位 true
    gameRunning = true

    //将问问题的次数清零
    askTimes = 0

    println("A new game has begun. Please ask your questions.")
}

//处理是否猜中数字的函数
func dealWithEqual(inputNum: Int64) {
    if (inputNum == guessedNumber) {
        println("#########################################")
        println("## Bingo!!! You guessed right!!!   ##")
        println("#########################################")
        println("Press n to start a new game and q to exit the game!")
        gameRunning = false
    } else {
        println("Sorry, you guessed wrong!")
    }
}

//根据操作符号和数字执行操作
func dealOpAndNum(opNum: OpAndNum) {
    askTimes++
    match (opNum.op) {
        case "==" | "=" => dealWithEqual(opNum.num)
        case ">" => PrintJudgmentResults(guessedNumber > opNum.num)
        case ">=" => PrintJudgmentResults(guessedNumber >= opNum.num)
        case "<" => PrintJudgmentResults(guessedNumber < opNum.num)
        case "<=" => PrintJudgmentResults(guessedNumber <= opNum.num)
        case _ => println("No")
    }
}

//根据判断结果打印判断信息
```

```
func PrintJudgmentResults(result: Bool) {
    if (result) {
        println("Yes! You guessed ${askTimes} times!")
    } else {
        println("No! You guessed ${askTimes} times!")
    }
}

main() {
    println("################################################################")
    println("## Welcome to the Figure guessing game!                       ##")
    println("## Each game produces a random number from 0 to 100           ##")
    println("## You can only use >,>= , == ,<,<= symbols and numbers       ##")
    println("## to ask me questions. Like > 50,<= 38, == 47 etc.           ##")
    println("## I can only answer yes or no!                               ##")
    println("## Press n to start a new game and q to exit the game!        ##")
    println("################################################################")

    //接收游戏玩家的输入并去除空白字符,输入异常就把输入当作q处理
    var input = (Console.stdIn.readln() ?? "q").trimAscii()

    //判断是否退出,不是退出就继续进行
    while (input != "q") {
        //是否开局新游戏
        if (input == "n" || input == "N") {
            startNewGame()
        } else {
            if (gameRunning) {
                //从输入提取操作符号和数字
                let opNum = getOpAndNum(input)

                match (opNum) {
                    //提取成功,执行操作
                    case Some(value) => dealOpAndNum(value)

                    //提取失败
                    case None => println("Your input is illegal!")
                }
            } else {
                println("Please start the game first!")
            }
        }

        //接收新的输入
        input = (Console.stdIn.readln() ?? "q").trimAscii()
    }
}
```

编译后运行该示例,可能的输出如下:

```
###########################################################
## Welcome to the figure guessing game!                  ##
## Each game produces a random number from 0 to 100      ##
## You can only use >,>= , == ,<,<=  symbols and numbers ##
## to ask me questions. Like >50,<=38,==47 etc.          ##
## I can only answer yes or no!                          ##
## Press n to start a new game and q to exit the game!   ##
###########################################################
n
A new game has begun. Please ask your questions.
>=50
No! You guessed 1 times!
<=25
Yes! You guessed 2 times!
>=12
No! You guessed 3 times!
<6
No! You guessed 4 times!
>9
No! You guessed 5 times!
>=7
Yes! You guessed 6 times!
8
Sorry, you guessed wrong!
7
###########################################################
## Bingo!!! You guessed right!!! ##
###########################################################
Press n to start a new game and q to exit the game!
q
```

# 第 21 章 高级集合类型

仓颉语言的集合类型除了第 15 章介绍的 Array 和 ArrayLis 外,还包括 HashSet 和 HashMap,本章将介绍这两种类型的用法。因为 HashSet 和 HashMap 都用到了 Hashable,下面首先介绍这个接口。

## 21.1 Hashable 接口

Hashable 接口的定义如下:

```
public interface Hashable {
/*
 * 获得实例类型的哈希值
 * 返回值 UInt64 - 返回实例类型的哈希值
 */
func hashCode(): UInt64
}
```

该接口声明了一个成员函数 hashCode(),实现了该接口的对象实例会根据本身的特征生成一个哈希值,不同实例生成的哈希值基本是不同的(也有可能相同,这种概率很小,和具体的实现算法有关)。

## 21.2 HashSet 集合

### 21.2.1 HashSet 的定义

HashSet 表示不重复元素的泛型集合,使用 HashSet < T >表示 HashSet 类型,其中,T 表示 HashSet 的元素类型,T 必须是实现了 Hashable 接口的类型。HashSet 位于 std 模块的 collection 包下,导入方式如下:

```
from std import collection.*
```

典型的 HashSet 定义,示例代码如下:

```
let hsInt: HashSet<Int64> = HashSet<Int64>()
var hsString = HashSet<String>(["cangjie","is","a","language"])
```

在创建实例时,元素作为初始化参数被传递给构造函数,如果有多个元素,则元素之间使用逗号分隔,外层使用方括号[]包裹。允许定义没有元素的空 HashSet。

### 21.2.2 访问 HashSet

#### 1. 访问成员

HashSet 不支持通过下标对单个元素进行访问,可以通过 for-in 循环遍历所有的元素。因为 HashSet 元素的插入位置和元素的哈希值有关,所以插入的顺序和遍历的顺序有可能不同,示例代码如下:

```
//Chapter21/HashSetDemo.cj

from std import collection.HashSet

main() {
    var hsString = HashSet<String>(["cangjie", "is", "the", "best", "programming", "language!"])

    //遍历 HashSet 的每个元素
    for (item in hsString) {
        println(item)
    }
}
```

编译后运行该示例,可能的输出如下:

```
cangjie
is
the
best
programming
language!
```

#### 2. 判断元素是否存在

因为 HashSet 主要用于存储不重复的元素,所以它的一个重要用法就是判断集合中是否存在特定的元素,对应的函数有以下两个。

- public func contains(element: T): Bool
- public func containsAll(elements: Collection<T>): Bool

第 1 个函数用于判断单个元素 element 是否存在,第 2 个函数用于判断 elements 列表里所有的元素是否都存在。

### 21.2.3 修改 HashSet

HashSet 支持对元素的添加、删除操作,可以每次操作一个元素,也可以进行批量操作,因为 HashSet 是引用类型,所有指向同一个实例的引用共享同样的元素。HashSet 主要的修改函数如下。

- public func put(element: T): Bool

put 函数添加单个元素 element 到 HashSet 中,如果 HashSet 不包含该元素,则添加该元素到 HashSet 并返回值为 true;如果 HashSet 中已存在该元素,则添加失败并返回值为 false。

- public funcputAll(elements: Collection<T>): Unit

putAll 函数批量添加 elements 中的多个元素到 HashSet,对于 HashSet 中不存在的元素,添加到 HashSet,已存在的元素不添加。

- public func remove(element: T): Bool

如果元素 element 存在于 HashSet 中,则将其移除,成功返回值为 true,失败返回值为 false。

- public func removeAll(elements: Collection<T>): Unit

从 HashSet 中移除所有存在于 elements 中的元素。

- public func clear(): Unit

移除 HashSet 中的所有元素。

- public func retainAll(elements: Collection<T>): Unit

保留 HashSet 中也存在于 elements 中的所有元素,其余的移除。

HashSet 操作的示例代码如下:

```
//Chapter21/manager_hashset_demo.cj

from std import collection.HashSet
from std import console.*
from std import random.Random
from std import convert.*

//处理用户输入
func dealWithInput(hsInt: HashSet<Int64>, input: String) {
    //如果是 c 命令,则清空 hsInt
    if (input.equals("c")) {
        hsInt.clear()
    } else if (input.equals("p")) {
        //如果是 p 命令,则打印 hsInt 所有的元素
        var outPrint = ""
        for (item in hsInt) {
            outPrint = outPrint.concat(item.toString() + ",")
        }
        outPrint = outPrint.trimRight(",")
        println(outPrint)
    }

    //如果输入长度大于1,说明可能是 a 或者 r 指令
```

```
        if (input.size > 1) {
            //如果是 a 或者 r 指令
            if (input[0] == 'a' || input[0] == 'r') {
                //得到命令后的数字
                let numOption = Int64.tryParse(input[1..])
                match (numOption) {
                    //数字解析成功
                    case Some(value) =>
                        //如果是添加命令,则将数字添加到 hsInt
                        if (input[0] == 'a') {
                            hsInt.put(value)
                        } else {
                            //否则就是移除命令,从 hsInt 移除数字
                            if (!hsInt.remove(value)) {
                                println("Removal failed")
                            }
                        }

                    //数字解析失败
                    case None => println("Parsing failed")
                }
            }
        }
    }
}

main() {
    //定义 HashSet 变量
    let hsInt: HashSet<Int64> = HashSet<Int64>()

    //随机生成 5 个整数并添加到 hsInt 集合中
    let rand = Random()
    var index = 0
    while (index < 5) {
        hsInt.put(rand.nextInt64(20))
        index++
    }

    //打印命令说明
    println("Please enter the operation command:")
    println("a + num:add num to the HashSet;r + num:remove num from the HashSet;")
    println("c:clear the HashSet;p:print all the elements;q:exit.")

    //接收用户输入并去除结尾的空白字符,输入异常就把输入内容置成"q"
    var input = (Console.stdIn.readln() ?? "q").trimAscii()

    //如果是 q 就退出
    while (input != "q") {
        //处理用户输入
        dealWithInput(hsInt, input)
```

```
            //接收用户输入
            input = (Console.stdIn.readln() ?? "q").trimAscii()
    }
}
```

在上述示例中，程序接收用户的输入指令，包括 a、r、c、p、q 等，根据指令执行对 HashSet 对象的操作，编译后运行该示例，可能的输入和输出如下：

```
Please enter the operation command:
a+num:add num to the HashSet;r+num:remove num from the HashSet;
c:clear the HashSet;p:print all the elements;q:exit.
p
16,18,4,8,12
r4
r8
a10
a11
p
16,18,10,11,12
q
```

## 21.2.4　HashSet 的容量和元素个数

HashSet 在实际实现时，会在内部维护一个数组，元素存放在数组里，存放的位置由元素的哈希值决定。内部数组的大小称为 HashSet 的容量，通过如下函数获取：

```
public func capacity(): Int64
```

HashSet 实际存放的元素个数通过另一个属性 size 获取，定义如下：

```
public prop size: Int64
```

在实例化 HashSet 时，会同时生成内部的数组，这个数组的容量在当前版本的仓颉语言中一般固定为 16，在将元素添加到 HashSet 时，理论上会出现添加的元素数量大于 HashSet 容量的情况，这时 HashSet 会对内部数组动态地扩容，在当前版本，扩容的容量为原先的 1.5 倍。因为扩容的过程有一定的资源和时间要求，所以最好事先估计出需要的 HashSet 容量，在定义时就初始化好合适的容量，避免后期频繁地动态扩容，可以通过如下的构造函数初始化指定容量的 HashSet：

```
public init(capacity: Int64)
```

其中，capacity 是初始化的容量。

下面通过具体的示例演示 HashSet 元素数量和容量的变化情况，代码如下：

```
//Chapter21/hashset_capacity_demo.cj

from std import collection.HashSet
```

```
main() {
    //定义 HashSet 变量
    let hsInt: HashSet<Int64> = HashSet<Int64>()

    //打印 HashSet 变量的初始容量
    println(hsInt.capacity())

    //将 30 个整数添加到 hsInt 集合中
    var index = 0
    while (index < 30) {
        //将 index 添加到 hsInt 集合中
        hsInt.put(index)

        //打印元素数量和 hsInt 容量对比信息
        println("${hsInt.size}:${hsInt.capacity()}")
        index++
    }
}
```

编译后运行该示例,输出如下:

```
16
1:16
2:16
3:16
4:16
5:16
6:16
7:16
8:16
9:16
10:16
11:16
12:16
13:16
14:16
15:16
16:16
17:24
18:24
19:24
20:24
21:24
22:24
23:24
24:24
25:36
26:36
27:36
28:36
```

```
29:36
30:36
```

除了可以在初始化时分配合适的容量外,也可以通过函数 reserve() 预留指定的容量,代码如下:

```
public func reserve(additional: Int64): Unit
```

将 HashSet 扩容 additional 指定的大小;当 additional 小于或等于 0 时,不发生扩容,当 HashSet 剩余容量大于或等于 additional 时,不发生扩容;当 HashSet 剩余容量小于 additional 时,取(原始容量的 1.5 倍向下取整)与(additional+已使用容量)中的最大值进行扩容。

## 21.3 HashMap 集合

### 21.3.1 HashMap 的定义

HashMap 表示键-值对元素的泛型集合,它的每个元素都是一个键-值对,使用 HashMap<K,V>表示 HashMap 类型,其中 K 表示 HashMap 的键类型,K 必须是实现了 Hashable 接口的类型;V 表示 HashMap 的值类型,V 可以是任意类型。HashMap 也是一种哈希表,表中的每个元素使用其键作为标识,可以通过键访问映射到的值。HashMap 位于 std 模块的 collection 包下,导入方式如下:

```
from std import collection.*
```

典型的 HashMap 定义的示例代码如下:

```
let dict: HashMap<Int64,String> = HashMap<Int64,String>()
let score = HashMap<String,Float64>([("001",76.5),("002",90.0)])
```

在创建实例时,包裹在小括号内的键-值对作为初始化参数被传递给构造函数,如果有多个键-值对,键-值对之间使用逗号分隔,外层使用方括号[]包裹。允许定义没有元素的空 HashMap。

### 21.3.2 访问 HashMap

HashMap 可以通过 for-in 循环遍历所有的元素,因为 HashMap 元素的插入位置和元素键的哈希值有关,所以插入的顺序和遍历的顺序有可能不同。HashMap 不支持通过序号作为下标索引对单个元素进行访问,但是 HashMap 支持通过元素的键作为下标索引访问映射元素的值,如果作为下标的键不存在,则会触发运行时异常,更好的方式是通过函数 contains() 判断给定的键是否存在,确认存在后再通过该键作为下标访问,示例代码如下:

```
//Chapter21/hashmap_demo.cj

from std import collection.HashMap

main() {
    let scoreMap = HashMap<String, Float64>([
        ("张飞", 95.5),
        ("李广", 68.0),
        ("王阳明", 100.0),
        ("赵明", 89.0)
    ])
    for ((name, score) in scoreMap) {
        println("${name}:${score}")
    }
    if (scoreMap.contains("赵明")) {
        println("赵明的成绩为 ${scoreMap["赵明"]}")
    }
}
```

编译后运行该示例,可能的输出如下:

```
张飞:95.500000
李广:68.000000
王阳明:100.000000
赵明:89.000000
赵明的成绩为 89.000000
```

HashMap 也支持通过函数方式获取指定键映射的值,函数定义如下:

```
public func get(key: K): Option<V>
```

该函数用于返回指定键 key 映射的值,如果 HashMap 不包含键 key,则返回 Option<V>.None。

### 21.3.3 修改 HashMap

HashMap 是一种可变的引用类型,提供了修改元素、添加元素、删除元素的功能。修改元素时,可以通过下标语法修改指定键映射的值,示例代码如下:

```
let score = HashMap<String,Float64>([("001",76.5),("002",90.0)])
score["002"] = 100.0
```

除了下标语法,HashMap 还有丰富的修改函数,常用的函数如下。
- public func put(key: K, value: V): Option<V>

将值 value 映射到键 key,并返回 key 原先映射的值。如果 key 在 HashMap 中已存在,则会用新的值 value 替换原来的值,并将原来的值使用 Option 封装返回;如果 key 在 HashMap 中不存在,则将键 key 和值 value 建立起映射关系后返回 Option.None。该函数的示例代码如下:

```
//Chapter21/hashmap_put_demo.cj

from std import collection.HashMap

main() {
    //定义 scoreMap 变量
    let scoreMap = HashMap<String, Float64>([("张飞", 95.5), ("李广", 68.0)])

    //添加关联的键-值对,因为原先没有键"赵明",所以返回值应该是 Option.None
    var optValue = scoreMap.put("赵明", 89.0)

    match (optValue) {
        case Some(value) => println("The original value was ${value}.")
        case None => println("There was no value.")
    }

    //添加关联的键-值对,因为原先有键"赵明",所以返回值应该是原先的值
    optValue = scoreMap.put("赵明", 99.0)

    match (optValue) {
        case Some(value) => println("The original value was ${value}.")
        case None => println("There was no value.")
    }
}
```

编译后运行该示例,输出如下:

```
There was no value.
The original value was 89.000000.
```

- public func putAll(elements: Collection<K,V>): Unit

批量添加映射关系,该函数会遍历列表 elements,把每个元素的键和值的映射关系添加到 HashMap,如果原先包含键的映射,则会使用新值替换原先的旧值。

- public func remove(key: K): Option<V>

如果 HashMap 存在指定的键,就删除指定键的映射,并返回使用 Option 封装的值;如果 HashMap 不存在指定的键,就返回 Option.None。

- public func removeAll(keys: Collection<K>): Unit

从 HashMap 中删除指定集合的映射(遍历 keys,如果对应的键在 HashMap 中存在,就删除该键和映射的值,如果不存在就忽略该键)。

- public func clear(): Unit

清除所有键-值对。

### 21.3.4 其他常用函数

- public func size: Int64

返回键-值对的个数。

- public func capacity(): Int64

返回 HashMap 的容量。

- public func reserve(additional: Int64): Unit

将 HashMap 扩容 additional 指定的大小；当 additional 小于或等于 0 时，不发生扩容；当 HashMap 剩余容量大于或等于 additional 时，不发生扩容；当 HashMap 剩余容量小于 additional 时，取（原始容量的 1.5 倍向下取整）与（additional＋已使用容量）中的最大值进行扩容。

- public func keys(): Keys<K>

返回 HashMap 中所有的键，并将所有的键存储在一个 Keys 容器中。

- public func values(): Values<V>

返回 HashMap 中包含的值，并将所有的值存储在一个 Values 容器中。

- public func isEmpty(): Bool

判断 HashMap 是否为空，如果是，则返回值为 true，否则返回值为 false。

### 21.3.5 综合应用示例

HashMap 是软件开发中应用非常广泛的一种类型，这里通过一个综合的示例，演示 HashMap 的常用用法。在这个示例中，会给出一段英文，示例程序会遍历该段英文，从中提取英文单词，并统计每个单词出现的次数，示例代码如下：

```
//Chapter21/word_statistics.cj

from std import collection.HashMap

//将单词 word 的小写形式加入 wordAndCountMap 中
//如果 word 已存在,则把该单词出现的数量加 1,否则把该单词对应的数量置为 1
func addWordToWordCountMap(word: String, wordAndCountMap: HashMap<String, Int64>) {
    let lowWord = word.toAsciiLower()
    
    //判断该单词的小写形式是否存在,如果存在,则把单词出现的数量加 1,否则把该单词对应的
    //数量置为 1
    if (wordAndCountMap.contains(lowWord)) {
        wordAndCountMap[lowWord] = wordAndCountMap[lowWord] + 1
    } else {
        wordAndCountMap[lowWord] = 1
    }
}

main() {
    let content = """
    I have a dream that one day this nation will rise up and live out the true meaning of its creed:
    "We hold these truths to be self - evident, that all men are created equal."
    I have a dream that one day on the red hills of Georgia, the sons of former slaves and the sons
    of former slave owners will be able to sit down together at the table of brotherhood."""
```

```
//这段话的总字符数量
let size = content.size

//从 0 开始的序号
var index = 0

//存储当前要处理的单词
var currentWord = ""

//存储单词和出现数量映射的 HashMap
let wordAndCountMap = HashMap<String, Int64>()

//遍历 content 的每个字符,从中提取单词并计数
while (index < size) {
    //当前要处理的字符
    let currentChar = content[index]

    //如果是 Ascii 字符,则追加到 currentWord 后面
    if (currentChar.isAsciiLetter()) {
        currentWord = currentWord.concat(currentChar.toString())
    } else {
        //否则就认为是单词分隔符

        //如果当前单词不为空,则加入 wordAndCountMap 中
        if (currentWord.size > 0) {
            addWordToWordCountMap(currentWord, wordAndCountMap)

            //把当前单词置空,准备生成下一个单词
            currentWord = ""
        }
    }
    index++
}

//处理最后一个字符是 Ascii 字符的情况,如果当前单词不为空,则加入 wordAndCountMap 中
if (currentWord.size > 0) {
    addWordToWordCountMap(currentWord, wordAndCountMap)
}

//输出单词个数
println("The number of words is ${wordAndCountMap.keys().size}")

//输出每个单词出现的数量
for ((word, count) in wordAndCountMap) {
    println("${word}:${count}")
}
}
```

编译后运行该示例，输出如下：

```
The number of words is 50
i:2
have:2
a:2
dream:2
that:3
one:2
day:2
this:1
nation:1
will:2
rise:1
up:1
and:2
live:1
out:1
the:5
true:1
meaning:1
of:5
its:1
creed:1
we:1
hold:1
these:1
truths:1
to:2
be:2
self:1
evident:1
all:1
men:1
are:1
created:1
equal:1
on:1
red:1
hills:1
georgia:1
sons:2
former:2
slaves:1
slave:1
owners:1
able:1
sit:1
down:1
together:1
at:1
table:1
brotherhood:1
```

# 第 22 章 模式匹配

## 22.1 match 表达式

### 22.1.1 pattern guard

在 6.3 节"match 表达式"里讲解了 match 表达式的基本用法,其中一种用法是包含待匹配值的 match 表达式,这种表达式允许在模式之后有一个可选的额外匹配条件(pattern guard),在同时匹配模式和条件的情况下,才会执行=>之后的代码,示例代码如下:

```
//Chapter22/pattern_guard_demo.cj

main() {
    printScore(Some(100))
    printScore(Some(65))
    printScore(Some(39))
    printScore(Option<Int64>.None)
}

//打印成绩的函数
func printScore(score: Option<Int64>): Unit {
    match (score) {
        //成绩大于或等于 90
        case Some(value) where (value >= 90) => println("Excellent!")

        //成绩大于或等于 80
        case Some(value) where (value >= 80) => println("Good!")

        //成绩大于或等于 60
        case Some(value) where (value >= 60) => println("Pass!")

        //成绩小于 60
        case Some(value) where (value < 60) => println("Fail!")

        //成绩为 None
```

```
            case None => println("No score!")

            //覆盖其他情况
            case _ => println("Illegal!")
    }
}
```

编译后运行该示例,输出如下:

```
Excellent!
Pass!
Fail!
No score!
```

### 22.1.2　match 表达式类型

match 表达式也有类型,如果上下文明确指定了 match 表达式的类型,就要求每个 case 分支中=>之后的代码块的类型是上下文所要求的类型的子类型;如果没有明确的类型要求,match 表达式的类型则由每个 case 分支中=>后的代码块的最小公共类型决定。match 表达式的类型的示例代码如下:

```
//Chapter22/match_type_demo.cj

main() {
    //获取成绩对应的评语
    var comment = getCommentOfScore(Some(65))
    println(comment)

    comment = getCommentOfScore(Option<Int64>.None)
    println(comment)

    //获取成绩数值
    let score = getScore(Some(80))
    println("The score is ${score}")
}

//获取成绩评语的函数
func getCommentOfScore(score: Option<Int64>): String {
    let comment: String = match (score) {
        //成绩大于或等于 90
        case Some(value) where (value >= 90) => "Excellent!"

        //成绩大于或等于 80
        case Some(value) where (value >= 80) => "Good!"

        //成绩大于或等于 60
        case Some(value) where (value >= 60) => "Pass!"
```

```
            //成绩小于60
            case Some(value) where (value < 60) => "Fail!"

            //成绩为None
            case None => "No score!"

            //覆盖其他情况
            case _ => "Illegal!"
        }
        return comment
    }

//有成绩时为实际成绩,无成绩时为-1
func getScore(score: Option<Int64>) {
    match (score) {
        case Some(value) => value
        case None => -1
    }
}
```

编译后运行该示例,输出如下:

```
Pass!
No score!
The score is 80
```

在上述示例中,函数 getCommentOfScore() 中的 match 表达式明确要求是 String 类型,所以每个 case 分支后的代码执行结果也是 String 类型;函数 getScore() 中的 match 表达式无明确的类型要求,但是每个 case 分支的代码所执行的结果都是 Int64 类型的,所以 match 表达式也是 Int64 类型的。

## 22.2 模式

对于包含匹配值的 match 表达式,case 后支持的模式有以下几种,分别是常量模式、通配符模式、变量模式、tuple 模式、类型模式和 enum 模式。

### 22.2.1 常量模式

常量模式使用字面量、let 定义的不可变变量及无参的枚举构造器作为模式,其中字面量可以是整数字面量、浮点数字面量、字符类型字面量、布尔类型字面量、字符串字面量(不支持插值)、Unit 字面量;匹配成功的条件是待匹配的值与常量模式表示的值相等;一个 case 后支持多个使用符号 | 连接的常量模式,匹配其中任何一个常量值就表示此 case 分支匹配成功。常量模式的示例代码如下:

```
//Chapter22/constant_pattern_demo.cj

main() {
    //打印成绩评价
    println(getScoreEvaluation('A'))
    println(getScoreEvaluation('C'))
}

//获取成绩评价,成绩分为 A、B、C、D、E 共 5 个等级
func getScoreEvaluation(score: Char): String {
    match (score) {
        case 'A' => "Excellent"
        case 'B' => "Good"
        case 'C' => "Pass"
        //多个常量并列作为模式
        case 'D' | 'E' => "Need to study hard"
        //通配符模式
        case _ => "Invalid score"
    }
}
```

编译后运行该示例,输出如下:

```
Excellent
Pass
```

## 22.2.2 通配符模式

通配符模式使用下画线表示,可以匹配任何值,一般放在最后一个 case 分支,用来匹配其他 case 分支不能覆盖的情况。

## 22.2.3 变量模式

变量模式是指在 case 后通过一个合法的标识符来匹配任何值,该标识符可以认为是一个不可变的变量,匹配后的值会和该变量绑定,在 => 之后可以通过变量访问绑定的值,示例代码如下:

```
//Chapter22/var_pattern_demo.cj

main() {
    let score = 'D'

    match (score) {
        case 'A' | 'B' => println("Your score is very good!")
        case value => println("Your score is ${value} and you need to study hard!")
    }
}
```

编译后运行该示例,输出如下:

```
Your score is D and you need to study hard!
```

### 22.2.4　元组模式

元组模式用于对元组值进行匹配,因为元组是多种类型的组合,所以它每个位置的元素都可以使用独立的模式,当元组每个位置的值都和对应位置的模式匹配时,才认为该 case 分支匹配成功,示例代码如下:

```
//Chapter22/tuple_pattern_demo.cj

main() {
    matchPerson(("山东", "王伟", 'A'))
    matchPerson(("江苏", "赵云", 'B'))
    matchPerson(("海南", "王斌", 'C'))
}

func matchPerson(person: (String,String,Char)): Unit {
    match (person) {
        case ("山东", name, score) => println(" ${name}来自山东,成绩是 ${score}")
        case (_, name, 'B') => println(" ${name}的成绩是 B")
        case (province, name, score) => println(" ${province} ${name}的成绩是 ${score}")
    }
}
```

编译后运行该示例,输出如下:

```
王伟来自山东,成绩是 A
赵云的成绩是 B
海南王斌的成绩是 C
```

### 22.2.5　类型模式

类型模式用来判断一个值的运行时类型是不是某种类型的子类型。类型模式有两种形式:_: Type 和 id: Type,对于第 1 种形式,会判断待匹配值的运行时类型是不是类型 Type 的子类型,如果是就匹配成功,否则匹配失败;对于第 2 种形式,首先判断待匹配值的运行时类型是不是类型 Type 的子类型,如果不是就匹配失败,如果是就匹配成功,然后把待匹配值转换为 Type 类型并和 id 表示的不可变变量绑定。类型模式的示例代码如下:

```
//Chapter22/type_pattern_demo.cj

from std import collection.*

//定义接口
interface Figure {}

//实现了接口的 Cirle 类
```

```
open class Circle <: Figure {
    Circle(let radius: Float64) {}
}

//实现了接口的 Cube 类
class Cube <: Figure {
    Cube(let length: Float64) {}
}

//继承自 Circle 的类
class ColorCircle <: Circle {
    ColorCircle(radius: Float64, let color: String) {
        super(radius)
    }
}

main() {
    matchType(ColorCircle(2.0, "red"))
    matchType(Cube(1.0))
    matchType(Circle(3.0))
    matchType(ArrayList<String>())
}

//匹配类型
func matchType(item: Object) {
    match (item) {
        case colorCircle: ColorCircle => println("The type is ColorCircle and the color is ${colorCircle.color}.")
        case cube: Cube => println("The type is Cube and the length is ${cube.length}.")
        case _: Figure => println("The type is Figure.")
        case _ => println("It\'s an object.")
    }
}
```

编译后运行该示例,输出如下:

```
The type is ColorCircle and the color is red.
The type is Cube and the length is 1.000000.
The type is Figure.
It's an object.
```

## 22.2.6 枚举模式

枚举模式用来匹配枚举类型(enum)的实例,枚举模式支持无参构造器和有参构造器,其中有参构造器的参数可以是常量也可以是变量,当是常量时,要求待匹配值和模式的构造器名称相同并且每个位置的参数值也相同;当是变量时,只要求构造器名称和参数个数相同,匹配成功时,对应位置的参数值会和变量绑定。枚举模式的示例代码如下:

```
//Chapter22/enum_pattern_demo.cj

//定义包含多种构造器的枚举
enum Color {
    Red | Green | Blue | Red(UInt8) | Green(UInt8) | Blue(UInt8) | Full(UInt8, UInt8, UInt8)
}

main() {
    matchColor(Color.Full(100, 100, 100))
    matchColor(Color.Red(100))
    matchColor(Color.Green(100))
    matchColor(Color.Blue)
}

//根据给定的颜色变量匹配对应的分支并打印信息
func matchColor(color: Color) {
    match (color) {
        case Full(red, green, blue) => println("Red: ${red};Green: ${green};Blue: ${blue}.")
        case Red(value)   => println("Red: ${value}.")
        case Green(value) => println("Green: ${value}.")
        case Blue(value)  => println("Blue: ${value}.")
        case Red   => println("The color is red.")
        case Green => println("The color is green.")
        case Blue  => println("The color is blue.")
    }
}
```

编译后运行该示例,输出如下:

```
Red:100;Green:100;Blue:100.
Red:100.
Green:100.
The color is blue.
```

在上述示例中,表达式 Color.Full(100,100,100)匹配的是有 3 个参数的分支,匹配成功后把每个参数都和对应位置的变量绑定,在=>后面可以访问这些变量。其他表达式的匹配方式更简单,此处就不一一说明了。

# 高级篇

# 第 23 章 函数的高级用法

## 23.1 Lambda 表达式

7min

### 23.1.1 Lambda 表达式的定义

Lambda 表达式可以通过简洁的语法表达丰富的功能,是函数式编程的重要内容,在现代编程语言中得到了广泛的支持。仓颉语言中 Lambda 表达式的定义形式如下:

```
{p1: T1, …, pn: Tn => expressions | declarations}
```

其中,=>符号前的部分是 Lambda 表达式的参数列表,如果有多个参数,则参数之间使用逗号(,)分隔;每个参数的名称和类型之间使用冒号(:)连接;=>符号之后是 Lambda 表达式体,是一组表达式或声明序列。Lambda 表达式的示例代码如下:

```
let lam = {left: Int64, right: Int64 => left + right}
```

Lambda 表达式也可以没有参数,示例代码如下:

```
let print = { => println("Hello cangjie!")}
```

不管有没有参数,Lambda 表达式的=>符号都不可以省略,除非作为尾随 Lambda(尾随 Lambda 说明见 23.3.1 节),例如下面示例的倒数第 2 行代码:

```
func funcDemo(omitFunc: () -> Unit) {
    omitFunc()
}

main() {
    funcDemo({ => println("Hello cangjie!")})
    funcDemo {println("Hello cangjie!")}
}
```

如果编译器能够推断出 Lambda 表达式参数的类型,则参数类型可以省略,示例代码如下:

```
let typeLam: (Int64, Int64) -> Int64 = {left: Int64, right: Int64 => left + right}
```

函数变量 typeLam 有给定的函数类型,从中可以推断出两个参数的类型都是 Int64 类型,所以上述表达式可以简写,简写后的代码如下:

```
let typeLam: (Int64, Int64) -> Int64 = {left, right => left + right}
```

同样,当 Lambda 表达式作为函数调用表达式的实参使用时,其参数类型也可以根据函数的形参类型推断,示例代码如下:

```
//Chapter23/type_inference_demo.cj
func stringCommentPrint(printFunc: (String) -> Unit, value: String) {
    printFunc(value)
}

func Int64CommentPrint(printFunc: (Int64) -> Unit, value: Int64) {
    printFunc(value)
}

main() {
    stringCommentPrint({value => println("/* ${value} */")}, "100")
    Int64CommentPrint({value => println("/* ${value} */")}, 100)
}
```

编译后运行该示例,输出如下:

```
/* 100 */
/* 100 */
```

在上述示例中,虽然作为实参的 Lambda 表达式的形式相同,但是因为调用它们的函数形参不同,在对 Lambda 表达式进行类型推断时,第 1 个表达式的参数被推断为 String 类型,第 2 个表达式的参数被推断为 Int64 类型。

### 23.1.2 Lambda 表达式的返回值

Lambda 表达式不支持显式声明返回类型,可以通过上下文推断出来,如果无法推断,编译器则会报错,主要的类型推断方式有以下几种。

**1. 上下文明确指定了返回类型**

(1) 赋值给变量时,根据变量类型推断。
(2) 作为参数时,根据使用处函数的形参类型推断。
(3) 作为返回值时,根据使用处函数的返回值类型推断。

返回类型推断的示例代码如下:

```
//Chapter23/return_type_inference_demo.cj

//该函数要求第 1 个参数是函数类型,并且返回值为 Int64
```

```
func paramLambda(lamInc: (Int64) -> Int64, num: Int64) {
    lamInc(num)
}

//该函数要求返回值是函数类型,并且返回函数的返回值类型是 Int64
func returnTypeLambda(): (Int64) -> Int64 {
    //根据函数返回值的要求,该 Lambda 表达式的返回类型是 Int64
    return {a => a + 1}
}

main() {
    //定义函数类型变量,该变量要求返回的类型是 Int64
    let addFunc: (Int64, Int64) -> Int64

    //因为变量要求函数的返回类型是 Int64,故此 Lambda 表达式的返回类型是 Int64
    addFunc = {a, b => a + b}

    //paramLambda 要求参数返回值的类型是 Int64,故此 Lambda 表达式的返回类型是 Int64
    paramLambda({a => a + 1}, 10)
}
```

**2. 上下文未明确指定返回类型**

此时 => 右侧 Lambda 表达式体的类型会被视为 Lambda 表达式的返回类型,规则等同于函数类型,如下面的示例,=> 右侧为两个 Int64 类型的值相加,返回类型为 Int64 类型,代码如下:

```
let incFunc = {value: Int64 => value + 1}
```

### 23.1.3 Lambda 表达式的调用

Lambda 表达式支持立即调用,也可以赋值给变量后使用变量名调用,示例代码如下:

```
//Chapter23/lambda_call_demo.cj

main() {
    //Lambda 表达式直接调用
    let sum = {a: Int64, b: Int64 => a + b}(1, 2)
    println(sum)

    //Lambda 表达式赋值给变量
    let productLambda = {a: Int64, b: Int64 => a * b}

    //通过变量调用 Lambda 表达式
    let product = productLambda(2, 3)
    println(product)
}
```

编译后运行该示例,输出如下:

```
3
6
```

## 23.2 闭包

### 23.2.1 闭包的定义

首先通过一个示例演示什么是闭包,代码如下:

```
//Chapter23/closure_demo.cj

//定义计数器类
class Counter {
    Counter(var count: Int64) {}

    //对变量count进行累加的函数
    func inc() {
        count++
    }
}

//该函数返回嵌套函数counting
func getCounterFunc(): () -> Int64 {
    //使用let定义引用类型的局部变量counter
    let counter = Counter(0)

    //嵌套函数访问了全局函数的局部变量counter
    func counting(): Int64 {
        counter.inc()
        return counter.count
    }

    //返回嵌套函数counting
    return counting
}

main() {
    //得到getCounterFunc中定义的嵌套函数
    let funcCount = getCounterFunc()

    //调用函数funcCount 5次并打印返回值
    for (i in 0..5) {
        println(funcCount())
    }
}
```

编译后运行该示例,输出如下:

```
1
2
3
4
5
```

在上述示例中,首先定义了一个计数器类 Counter,然后在函数 getCounterFunc 中定义了一个局部函数 counting,需要特别注意的有两点:第一点,函数 counting 访问了上层函数中的局部变量 counter,并调用了它的 inc 方法,也就是说函数 counting 访问了不属于本函数的局部变量,或者让它捕获了该变量,这样,函数 counting 和它捕获的变量 counter 一起构成了一个闭包;第二点,外层函数 getCounterFunc 把局部函数 counting 作为返回值,这样可以把函数 counting 返回它的定义作用域之外。在示例的 main 函数中,局部函数 counting 被赋给了变量 funcCount,并调用了 5 次,可以看到函数可以正常运行,并输出了从 1 到 5 的数字。这说明,变量 counter 在脱离了它被定义的作用域后,仍然可以通过函数 counting 访问。在仓颉语言中,闭包的定义如下:"一个函数或 Lambda 表达式从定义它的静态作用域中捕获了变量,函数或 Lambda 和捕获的变量一起被称为一个闭包,这样即使脱离了闭包定义所在的作用域,闭包也能正常运行。"

## 23.2.2 捕获变量的状态

在对构成闭包的函数或者 Lambda 表达式进行调用时,被捕获的变量可以一直保持状态,在上述示例中,对 funcCount 的每一次调用最终都会触发捕获变量调用 inc 成员函数,也就是对类 Counter 的成员变量 count 进行累加,因为可以保持状态,count 的值也从 1 累加到了 5。

多次调用外层函数(如 getCounterFunc)返回局部函数(如 counting)形成的多个闭包是互相独立的,它们捕获的变量的状态变化互相不影响,把 23.2.1 节的示例代码的 main 函数修改一下,多次调用 getCounterFunc,形成多个闭包,修改后的代码如下:

```
//Chapter23/capture_variable_demo.cj

//定义计数器类
class Counter {
    Counter(var count: Int64) {}

    //对变量 count 进行累加的函数
    func inc() {
        count++
    }
}

//该函数返回嵌套函数 counting
func getCounterFunc(): () -> Int64 {
```

```
    //使用 let 定义引用类型的局部变量 counter
    let counter = Counter(0)

    //嵌套函数访问了全局函数的局部变量 counter
    func counting(): Int64 {
        counter.inc()
        return counter.count
    }

    //返回嵌套函数 counting
    return counting
}

main() {
    //使用 getCounterFunc 得到两个闭包
    let funcCount1 = getCounterFunc()
    let funcCount2 = getCounterFunc()

    //分别调用闭包中的函数 5 次并打印返回值
    for (i in 0..5) {
        println("funcCount1:${funcCount1()}")
        println("funcCount2:${funcCount2()}")
    }
}
```

编译后运行该示例，输出如下：

```
funcCount1:1
funcCount2:1
funcCount1:2
funcCount2:2
funcCount1:3
funcCount2:3
funcCount1:4
funcCount2:4
funcCount1:5
funcCount2:5
```

上述示例说明，每个闭包中捕获的变量都是独立的，对一个闭包变量的修改不会影响另一个闭包变量的状态。

### 23.2.3 可变变量的闭包

在 23.2.1 节和 23.2.2 节的示例中，捕获的变量为使用 let 关键字修饰的不可变变量，在仓颉语言中也可以捕获使用 var 修饰的可变变量，但是，为了防止这种类型的闭包逃逸，仓颉语言中规定捕获可变变量的闭包只能被调用，而不能作为第一类对象使用，包括不能赋值给变量，不能作为实参或返回值使用，也不能直接将闭包的名字作为表达式使用。下例代码中的 changeDemo 函数捕获了可变变量 demoStr，因为这个函数被当作了返回值，编译时会报错(changeDemo captured a mutable variable directly，changeDemo cannot be used as a return value)：

```
func getChangeDemoFunc(): (String) -> Unit {
    var demoStr = "demo"

    func changeDemo(newStr: String): Unit {
        demoStr = newStr
    }

    return changeDemo
}

main() {
    let funcChange = getChangeDemoFunc()
    funcChange("newValue")
}
```

函数对变量的捕获具有传递性，示例代码如下：

```
func getChangeDemoFunc(): () -> Unit {
    var demoStr = "demo"

    func changeDemo(newStr: String): Unit {
        demoStr = newStr
    }

    func callClosureFunc() {
        changeDemo("")
    }
    return callClosureFunc
}
```

局部函数 changeDemo 捕获了外层函数 getChangeDemoFunc 的局部可变变量 demoStr，形成了一个闭包，同样作为局部函数的 callClosureFunc 虽然没有直接捕获变量，但是它调用了函数 changeDemo，根据变量捕获的传递性，callClosureFunc 也不能作为第一类对象使用，上述示例编译时会给出如下的错误提示：callClosureFunc captured a mutable variable transitively, callClosureFunc cannot be used as a return value。

## 23.3 函数调用语法糖

### 23.3.1 尾随闭包

当函数调用的最后一个实参是 Lambda 表达式时，可以将 Lambda 表达式放在函数调用的尾部，即小括号外面，这种语法被称为尾随闭包。尾随闭包可以增加语言的可扩展性，使代码更优雅，示例代码如下：

```
func fmtOutput(value: String, fmtFun: (String) -> String) {
    println(fmtFun(value))
}

main() {
    fmtOutput("hello", {item => "<b>${item}</b>"})
    fmtOutput("cangjie", {item => "<strong>${item}</strong>"})

    fmtOutput("hello") {item => "<b>${item}</b>"}

    fmtOutput("cangjie") {item => "<strong>${item}</strong>"}
}
```

在上述示例中,定义了一个全局函数 fmtOutput,它可接收两个参数,最后一个参数是函数类型,在实际传参时可以传递 Lambda 表达式;在 main 函数里,前两行是正常的函数调用,把 Lambda 表达式写到了函数的参数里,后面的代码使用尾随闭包的语法改写了函数的调用方式,使代码更简洁,更容易理解。

当函数调用有且只有一个 Lambda 表达式的实参时,还可以省略(),只写 Lambda 表达式,下面使用一个对 ArrayList 集合进行条件删除的示例进行演示。首先看一下 ArrayList 类型的 removeIf 函数,代码如下:

```
public func removeIf(predicate: (T) -> Bool): Unit
```

它通过传入的比较函数,对每个元素进行判断,对于返回值是 true 的元素进行删除,示例代码如下:

```
//Chapter23/trailing_closure_demo.cj

from std import random.Random
from std import collection.ArrayList

main() {
    //初始化随机数
    let rand = Random()

    let numList = ArrayList<Int64>()

    //随机生成 10 个整数并保存到 numList 列表中
    for (i in 0..10) {
        numList.append(rand.nextInt64(100))
    }

    //输出这 10 个数字
    println(numList)

    //先复制一份列表
    var cloneList = numList.clone()
```

```
    //对复制后的列表使用正常的函数调用方式删除列表中的偶数
    cloneList.removeIf({item: Int64 => item % 2 == 0})

    //输出剩下的数字
    println(cloneList)

    //再复制一份列表
    cloneList = numList.clone()

    //对复制后的列表使用尾随闭包的函数调用方式删除列表中的偶数
    cloneList.removeIf {item: Int64 => item % 2 == 0}

    //输出剩下的数字
    println(cloneList)
}
```

编译后运行该示例,可能的输出如下:

```
[74, 67, 45, 2, 99, 82, 41, 63, 72, 39]
[67, 45, 99, 41, 63, 39]
[67, 45, 99, 41, 63, 39]
```

## 23.3.2 管道表达式

在需要对输入数据进行一系列的处理时,可以使用管道(pipeline)表达式简化代码的编写,管道表达式的语法如下:

```
p |> f
```

该语法等价于如下形式的语法糖:

```
f(p)
```

其中,f 是函数类型的表达式,p 的类型是 f 的参数类型的子类型,示例代码如下:

```
//Chapter23/pipeline_demo.cj
//小麦
class Wheat {}

//面粉
class Flour {}

//面团
class Dough {}

//馒头
class SteamedBun {}

//磨
```

```
func smash(wheat: Wheat): Flour {
    Flour()
}

//揉面
func knead(flour: Flour): Dough {
    Dough()
}

//蒸制
func steame(Dough: Dough): SteamedBun {
    SteamedBun()
}

main() {
    //小麦
    let wheat = Wheat()

    //小麦加工成馒头的流程
    //小麦磨成面粉
    let flour = smash(wheat)

    //面粉做成面团
    let dough = knead(flour)

    //面团蒸成馒头
    let steamedBun = steame(dough)

    //使用管道语法,简化从小麦到馒头的处理过程
    //本行代码相当于上面多行代码的功能,而且更易懂
    let newSteamedBun = wheat |> smash |> knead |> steame
}
```

在上述示例里,使用代码模拟了小麦从磨成面粉、揉成面团到最终蒸成馒头的过程,如果使用通常的函数调用写法,需要多行完成,如果使用管道表达式,只需简洁直观的一行代码便可以完成。

管道表达式在使用时要注意,不能与无默认值的命名形参函数直接一同使用,因为无默认值的命名形参函数必须给出命名实参才可以调用,而管道表达式无法给出命名参数,此种情况可以通过 Lambda 表达式间接完成,示例代码如下:

```
func f(a!: Int64): Unit {}

var x = 1 |> {p: Int64 => f(a: p)}
```

## 23.3.3 组合操作符

组合(composition)表达式可以表示两个单参数函数的组合,语法如下:

```
f ~> g
```

该语法等价于如下形式的语法糖：

```
{x => g(f(x))}
```

其中，f、g 均为只有一个参数的函数类型的表达式，f 的返回类型是 g 的参数类型的子类型。组合操作符的示例代码如下：

```
//Chapter23/composition_demo.cj

//小麦
class Wheat {}

//面粉
class Flour {}

//面团
class Dough {}

//馒头
class SteamedBun {}

//磨
func smash(wheat: Wheat): Flour {
    Flour()
}

//揉面
func knead(flour: Flour): Dough {
    Dough()
}

//蒸制
func steame(Dough: Dough): SteamedBun {
    SteamedBun()
}

main() {
    //小麦
    let wheat = Wheat()

    //将小麦加工成馒头的流程组合成一个函数变量
    let compFunc = smash ~> knead ~> steame

    let comSteamedBun = compFunc(wheat)
}
```

上述示例与 23.3.2 节的示例类似，这里把多个操作的函数组合成一个函数变量，然后调用该函数变量就能达到原来调用多个函数的效果。

# 第 24 章 并发

现代的计算机大都使用多核处理器，并且呈现出核心数越来越多的趋势，通过对计算任务的并发处理，可以充分发挥这种优势。在具体的线程处理层面，有的语言直接调用操作系统 API 来创建线程，也就是每个线程对应一个操作系统线程，一般称为 1∶1 的线程模型；也有一些编程语言提供特殊的线程实现，允许多个线程在不同数量的操作系统线程的上下文中切换执行，这种被称为 M∶N 的线程模型，即 M 个线程在 N 个操作系统线程上调度执行，其中 M 和 N 不一定相等。

仓颉语言屏蔽了线程具体实现的底层细节，提供了一个统一、高效、用户友好的仓颉线程概念，采用的也是 M∶N 的线程模型，本质上是一个用户态的轻量级线程，支持抢占，并且相比操作系统线程内存资源占用更小。

## 24.1 仓颉线程

### 24.1.1 线程睡眠函数 sleep

在介绍具体的仓颉线程以前，先介绍 sleep 函数，它在本章的示例中应用较多。sleep 函数的定义如下：

```
public func sleep(dur: Duration): Unit
```

该函数会阻塞当前运行的线程 dur 时间，也就是说，当前运行的线程会睡眠 dur 时间，然后恢复执行（如果 dur <= Duration.Zero，则当前线程只会让出执行资源，并不会进入睡眠）。下面给出一个使用 sleep 函数的演示示例，该示例中主线程（程序启动时即被创建的线程）会被阻塞 5 次，每次会睡眠 1s，并在恢复后打印当前时间，代码如下：

```
//Chapter24/sleep_demo.cj

from std import time.*
from std import sync.*
```

```
main() {
    for (i in 0..5) {
        //本次调用休眠 1s
        sleep(Duration.second)
        println(DateTime.now().toString("yyyy-MM-dd HH:mm:ss"))
    }
}
```

编译后运行该示例,可能的输出如下:

```
2022-05-02 15:07:53
2022-05-02 15:07:54
2022-05-02 15:07:55
2022-05-02 15:07:56
2022-05-02 15:07:57
```

## 24.1.2　创建仓颉线程

要创建一个新的仓颉线程,可以使用关键字 spawn,后跟一个无形参的 Lambda 表达式,该 Lambda 表达式即为在新创建的线程中执行的代码,示例代码如下:

```
//Chapter24/spawn_demo.cj

from std import time.*
from std import sync.*

main() {
    //创建一个新的仓颉线程,在新线程里打印 5 次当前时间
    spawn {
        => for (i in 0..5) {
            sleep(Duration.second)
            println("New Thread ${i} ${DateTime.now()}")
        }
    }

    //在主线程里打印 3 次当前时间
    for (i in 0..3) {
        sleep(Duration.second)
        println("Main Thread ${i} ${DateTime.now()}")
    }
}
```

编译后运行该示例,可能的输出如下:

```
Main Thread 0 2023-09-12T13:59:04Z
New Thread 0 2023-09-12T13:59:04Z
Main Thread 1 2023-09-12T13:59:05Z
New Thread 1 2023-09-12T13:59:05Z
New Thread 2 2023-09-12T13:59:06Z
Main Thread 2 2023-09-12T13:59:06Z
```

从以上示例可以看出，新创建的子线程（被主线程创建的线程一般称为子线程）和主线程是同时运行的，当主线程执行完毕后主线程结束，虽然此时子线程还没执行完，但是子线程也随着主线程的结束而结束了。

### 24.1.3 等待线程结束并获取返回值

在24.1.2节的示例中，主线程和子线程同时执行，因为主线程结束了，所以子线程也被迫提前结束了，这样可能会导致程序产生不可预期的结果。为了解决这种问题，仓颉语言提供了Future<T>类型，可以作为spawn表达式的返回值，通过该返回值等待线程执行结束。Future<T>类型的主要成员函数如下所示。

1) public func get()：T

阻塞当前线程，等待当前Future对象对应的线程的结果。

2) public func get(ns：Int64)：Option<T>

阻塞当前线程，等待当前Future对象对应的线程的结果，如果相应的线程在纳秒内未完成执行，则该函数将返回None。

3) public func tryGet()：Option<T>

尝试获取结果，不会阻塞当前线程，如果相应的线程未执行完成，则该函数返回None，否则返回Some。

下面通过Future<T>类型改写24.1.2节的示例，使其确保子线程执行完毕，示例代码如下：

```
//Chapter24/future_demo.cj

from std import time.*
from std import sync.*

main(): Unit {
    //创建子线程并获取返回值，在新线程里打印5次当前时间
    let newThreadFuture = spawn {
        => for (i in 0..5) {
            sleep(Duration.second)
            println("New Thread ${i} ${DateTime.now()}")
        }
    }

    //在主线程里打印3次当前时间
    for (i in 0..3) {
        sleep(Duration.second)
        println("New Thread ${i} ${DateTime.now()}")
    }

    //等待子线程执行结束
    newThreadFuture.get()
}
```

编译后运行该示例，可能的输出如下：

```
New Thread 0 2023 - 09 - 12T14:30:06Z
New Thread 0 2023 - 09 - 12T14:30:06Z
New Thread 1 2023 - 09 - 12T14:30:07Z
New Thread 1 2023 - 09 - 12T14:30:07Z
New Thread 2 2023 - 09 - 12T14:30:08Z
New Thread 2 2023 - 09 - 12T14:30:08Z
New Thread 3 2023 - 09 - 12T14:30:09Z
New Thread 4 2023 - 09 - 12T14:30:10Z
```

在上述示例里，通过在代码的最后调用 newThreadFuture.get() 阻塞了主线程，直到子线程执行结束，从最后的输出可以看出，子线程和主线程的代码都执行完毕。

Future<T>类型的 func get()：T 函数除了可以阻塞等待线程执行结束，还可以得到线程执行的结果，示例代码如下。

```
//Chapter24/get_result_demo.cj

from std import random.Random

main(): Unit {
    //产生两个随机的整数
    let rand = Random()
    let leftValue = rand.nextInt64(100)
    let rightValue = rand.nextInt64(100)

    //创建子线程并获取随机数相加的返回值
    let newThreadFuture = spawn { => leftValue + rightValue}

    //等待子线程执行结束
    let addResult = newThreadFuture.get()

    println("${leftValue} + ${rightValue} = ${addResult}")
}
```

编译后运行该示例，可能的输出如下：

```
86 + 84 = 170
```

通过调用 get() 可以得到线程内 Lambda 表达式具体的返回值。

函数 get(ns：Int64)：Option<T>的示例代码如下：

```
//Chapter24/get_result_timeout_demo.cj

from std import random.Random
from std import time.*
from std import sync.*

main(): Unit {
    //产生两个随机的整数
```

```
    let rand = Random()
    let leftValue = rand.nextInt64(100)
    let rightValue = rand.nextInt64(100)

    //创建子线程并获取随机数相加的返回值
    let newThreadFuture = spawn {
        =>
        sleep(Duration.second * 10)        //等待 10s
        leftValue + rightValue
    }

    //等待子线程执行结束,最长等待 5s
    let addResult = newThreadFuture.get(5 * 1000 * 1000 * 1000)

    //解构 Result 中的值
    match (addResult) {
        case Some(value) => println("${leftValue} + ${rightValue} = ${value}")
        case None => println("Error")
    }
}
```

编译后运行该示例,输出如下:

```
Error
```

## 24.2 同步

### 24.2.1 数据竞争

在多线程场景下运行的程序,有可能会带来非预期的结果,下面通过一段代码,分别演示在主线程里对一个变量累加十万次与使用十个线程且每个线程都对该变量累加一万次的区别,代码如下:

```
//Chapter24/data_race_demo.cj

from std import collection.ArrayList
from std import sync.*
from std import time.*

//计数器类
class Counter {
    Counter(var count: Int64) {}

    //对变量 count 进行累加的函数
    func inc() {
        count++
    }
```

```
    //累加器清零
    func reset() {
        count = 0
    }
}

main() {
    //用来累加计数的变量
    let counter = Counter(0)

    //在主线程里累加 10 万次
    for (i in 0..10) {
        for (j in 0..10000) {
            //等待 1μs
            sleep(Duration.microsecond)

            //累加器加 1
            counter.inc()
        }
    }

    //打印累加器的值
    println(counter.count)

    //累加器清零
    counter.reset()

    //以下代码通过 10 个线程各累加 1 万次
    //存放线程执行结果的列表
    let resultList = ArrayList<Future<Unit>>()

    //循环 10 次,将创建 10 个线程
    for (i in 0..10) {
        //创建一个用来累加的线程,线程内循环一万次,每次对累加器加 1
        let newThreadFuture = spawn {
            => for (j in 0..10000) {
                //等待 1μs
                sleep(Duration.microsecond)

                //累加器加 1
                counter.inc()
            }
        }

        //将线程执行结果放入列表
        resultList.append(newThreadFuture)
    }
```

```
    //遍历线程执行结果列表,确保每个线程执行完毕
    for (result in resultList) {
        result.get()
    }

    //打印累加器的值
    println(counter.count)
}
```

编译并多次运行该示例,输出的结果可能如下:

```
root@instance-rwqyn1pd:/data/code/demo/src#./main
100000
99569
root@instance-rwqyn1pd:/data/code/demo/src#./main
100000
99820
root@instance-rwqyn1pd:/data/code/demo/src#./main
100000
99554
```

从上述示例可以看出,在主线程下,对一个变量从 0 开始累加 10 万次,得到的结果一定是 100 000,但是,如果使用多个线程来累加,则得到的结果总会少上几百,并且每次的累加值也是不确定的。

出现这种情况的原因是多个线程同时操作一个变量的结果,下面粗略模拟一下本示例中多线程操作变量的过程,分析累加结果出错的原因。在本示例中,累加的最终代码在计数器类 Counter 的成员函数 inc 中,也就是对成员变量 count 执行++操作,代码如下:

```
func inc() {
    count++
}
```

假设在某一个时刻,count 的值是 100:

(1) 线程 1 读取 count 的值为 100。
(2) 线程 2 读取 count 的值为 100。
(3) 线程 1 把 100 加 1,得到 101,并赋给 count,此时 count 为 101。
(4) 线程 2 把 100 加 1,得到 101,并赋给 count,此时 count 为 101。

从上述模拟过程可以看出,线程 1 和线程 2 都累加完毕,此时 count 的值为 101,而不是 102,出现这种情况的原因是变量 count 允许同时被多个线程读写,即使变量的值发生了变化,读取它变化前值的线程并没有被通知该变化,还是使用之前的值参与运算,这样就出现了不可预期的结果。

### 24.2.2 原子操作

在 24.2.1 节的示例中,执行变量累加的两个操作,也就是读取变量值和修改变量值的

操作是分开的,因为允许多个线程对共享的变量同时操作,导致出现数据竞争,如果把这个过程设计成原子的,也就是读取值和修改值的过程不可中断,就可以避免数据竞争的发生。在仓颉语言中,提供了原子操作 Atomic,可以对整数类型、Bool 类型和引用类型执行原子操作。对整数类型的原子操作支持基本的读写、交换及算术运算操作,包括 Int8、Int16、Int32、Int64、UInt8、UInt16、UInt32、UInt64 等类型,下面以 Int64 类型为例,介绍各个成员函数,Int64 类型对应的原子类型为 AtomicInt64,成员函数如下:

1. public init(val：Int64)

构造函数,实例的初始值为参数 val 的值。

2. public init(val：Int64, memoryOrder：MemoryOrder)

构造函数,通过参数 val 和 memoryOrder 指定的初始值和内存顺序,构造整数类型的原子操作实例。

MemoryOrder 是内存顺序的枚举类型,包括以下的成员:

- Relaxed：宽松顺序,不对执行顺序做保证;
- Consume：消费顺序,所有后续的有关本原子类型的操作,必须在本条原子操作完成之后执行;
- Acquire：获得顺序,所有后续的读操作必须在本条原子操作完成后执行;
- Release：释放顺序,所有之前的写操作完成后才能执行本条原子操作;
- AcqRel：释放获得顺序,同时包括 Acquire 和 Release;
- SeqCst：序列一致顺序,全部存取都按顺序执行。

3. public func load()：Int64

获取操作,采用默认内存排序方式,读取原子类型的值并返回。

4. public func load(memoryOrder：MemoryOrder)：Int64

获取操作,采用参数 memoryOrder 指定的内存排序方式,读取原子类型的值并返回。

5. public func store(val：Int64)：Unit

写入操作,采用默认内存排序方式,将参数 val 指定的值写入原子类型。

6. public func store(val：Int64, memoryOrder：MemoryOrder)：Unit

写入操作,采用参数 memoryOrder 指定的内存排序方式,将参数 val 指定的值写入原子类型。

7. public func swap(val：Int64)：Int64

交换操作,采用默认内存排序方式,将参数 val 指定的值写入原子类型,并返回写入前的值。

8. public func swap(val：Int64, memoryOrder：MemoryOrder)：Int64

交换操作,采用参数 memoryOrder 指定的内存排序方式,将参数 val 指定的值写入原子类型,并返回写入前的值。

9. public func compareAndSwap(old：Int64, new：Int64)：Bool

比较后交换操作,采用默认内存排序方式,比较当前原子类型的值与参数 old 指定的值是否相等：若相等，则写入参数 new 指定的值，并返回值为 true；否则不写入值，并返回值为 false。

10. public func compareAndSwap(old：Int64，new：Int64，memoryOrder1：MemoryOrder memoryOrder2：MemoryOrder)：Bool

比较后交换操作，采用参数 memoryOrder1 和 memoryOrder2 指定的内存排序方式，比较当前原子类型的值与参数 old 指定的值是否相等：若相等，则写入参数 new 指定的值，并返回值为 true；否则不写入值，并返回值为 false。

11. public func fetchAdd(val：Int64)：Int64

加操作，采用默认内存排序方式，将原子类型的值与参数 val 进行加操作，将结果写入当前原子类型实例，并返回加操作前的值。

12. public func fetchAdd(val：Int64，memoryOrder：MemoryOrder)：Int64

加操作，采用参数 memoryOrder 指定的内存排序方式，将原子类型的值与参数 val 进行加操作，将结果写入当前原子类型实例，并返回加法运算前的值。

13. public func fetchSub(val：Int64)：Int64

减操作，采用默认内存排序方式，将原子类型的值与参数 val 进行减操作，将结果写入当前原子类型实例，并返回减操作前的值。

14. public func fetchSub(val：Int64，memoryOrder：MemoryOrder)：Int64

减操作，采用参数 memoryOrder 指定的内存排序方式，将原子类型的值与参数 val 进行减操作，将结果写入当前原子类型实例，并返回减操作前的值。

15. public func fetchAnd(val：Int64)：Int64

与操作，采用默认内存排序方式，将当前原子类型实例的值与参数 val 进行与操作，将结果写入当前原子类型实例，并返回与操作前的值。

16. public func fetchAnd(val：Int64，memoryOrder：MemoryOrder)：Int64

与操作，采用参数 memoryOrder 指定的内存排序方式，将当前原子类型实例的值与参数 val 进行与操作，将结果写入当前原子类型实例，并返回与操作前的值。

17. public func fetchOr(val：Int64)：Int64

或操作，采用默认内存排序方式，将当前原子类型实例的值与参数 val 进行或操作，将结果写入当前原子类型实例，并返回或操作前的值。

18. public func fetchOr(val：Int64，memoryOrder：MemoryOrder)：Int64

或操作，采用参数 memoryOrder 指定的内存排序方式，将当前原子类型实例的值与参数 val 进行或操作，将结果写入当前原子类型实例，并返回或操作前的值。

19. public func fetchXor(val：Int64)：Int64

异或操作，采用默认内存排序方式，将当前原子类型实例的值与参数 val 进行异或操作，将结果写入当前原子类型实例，并返回异或操作前的值。

20. public func fetchXor(val：Int64，memoryOrder：MemoryOrder)：Int64

异或操作，采用参数 memoryOrder 指定的内存排序方式，将当前原子类型实例的值与参数 val 进行异或操作,将结果写入当前原子类型实例,并返回异或操作前的值。

使用原子操作,可以解决数据竞争问题,下面通过 AtomicInt64 类型,改写 24.2.1 节中的示例,测试一下是否可以解决累计结果出错的问题,示例代码如下:

```
//Chapter24/atomic_demo.cj

from std import collection.ArrayList
from std import sync.*
from std import time.*

main() {
    //用来累加计数的变量
    let counter = AtomicInt64(0)

    //以下代码通过10个线程各累加1万次
    //存放线程执行结果的列表
    let resultList = ArrayList<Future<Unit>>()

    //循环10次,将创建10个线程
    for (i in 0..10) {
        //创建一个用来累加的线程,线程内循环一万次,每次对累加器加1
        let newThreadFuture = spawn { => for (j in 0..10000) {
            //等待1μs
            sleep(Duration.microsecond)

            //累加器加1
            counter.fetchAdd(1)
        }}

        //将线程执行结果放入列表
        resultList.append(newThreadFuture)
    }

    //遍历线程执行结果列表,确保每个线程执行完毕
    for (result in resultList) {
        result.get()
    }

    //打印累加器的值
    println(counter.load())
}
```

编译并多次运行该示例,输出如下:

```
root@instance-rwqyn1pd:/data/code/demo/src#./main
100000
root@instance-rwqyn1pd:/data/code/demo/src#./main
100000
root@instance-rwqyn1pd:/data/code/demo/src#./main
100000
```

从上述示例可以看出，使用原子操作，较好地解决了多线程引起的数据竞争问题。

Bool 类型和引用类型的原子操作只支持读写和交换操作，如下所示。

(1) load：读取。

(2) store：写入。

(3) swap：交换，返回交换前的值。

(4) compareAndSwap：比较再交换，交换成功后的返回值为 true，否则返回值为 false。

其中，Bool 类型的原子操作类为 AtomicBool，引用类型的原子操作类为 AtomicReference，下面通过一个具体的示例演示引用类型的原子操作使用方法，代码如下：

```
//Chapter24/atomic_reference_demo.cj

from std import sync.AtomicReference

//用来演示的类
class Demo {
    Demo(let name: String) {}
}

main() {
    //创建 demo 实例
    let demo = Demo("demo")

    //使用 demo 初始化引用类型原子操作类实例 atomicRef
    let atomicRef = AtomicReference(demo)

    //创建新的 demo 实例 newDemo
    let newDemo = Demo("new")

    //把 newDemo 写入 atomicRef
    atomicRef.store(newDemo)

    //创建另一个 demo 实例 otherDemo
    let otherDemo = Demo("other")

    //使用 otherDemo 和 atomicRef 中的 demo 实例做对比，因为 atomicRef 中的实例是 newDemo
    //所以对比失败，交换也失败
    var swap = atomicRef.compareAndSwap(otherDemo, demo)

    //打印交换结果
    println(swap)

    //使用 newDemo 和 atomicRef 中的 demo 实例做对比，对比成功，把 demo 写入 atomicRef 中
    swap = atomicRef.compareAndSwap(newDemo, demo)

    //打印交换结果
    println(swap)
```

```
    //打印atomicRef中实例的成员变量name
    println(atomicRef.load().name)
}
```

编译后运行该示例,输出如下:

```
false
true
demo
```

## 24.2.3 互斥锁

原子操作类可以解决单个变量的数据竞争问题,在使用时需要把原始的数据类型转换为对应的原子操作类,稍微有点不便。在解决多线程数据竞争问题时,也可以考虑使用互斥锁,它可以保护一个或者多个变量免于数据竞争的风险。对于存在多线程数据竞争的代码段,一般称为临界区,互斥锁可以对临界区加以保护,任意时刻允许最多一个线程能够执行临界区的代码。当一个线程试图获取一个已被其他线程持有的锁时,该线程会被阻塞,直到锁被释放,该线程才会被唤醒。

如果某一个线程已经持有了一个互斥锁,则该线程在尝试重新获取同一个锁时,如果可以立即获得该锁,则这种锁被称为可重入互斥锁。可重入互斥锁要保证获取和释放锁的动作是一一对应的,这样才能最终释放锁。

提供可重入互斥锁功能的类是 ReentrantMutex,它的主要成员函数如下:

1. public init()

构造函数。

2. public func lock(): Unit

锁定互斥体,如果互斥体已被锁定,则阻塞。

3. public func tryLock(): Bool

尝试锁定互斥体,如果互斥体已被锁定,则返回值为 false,否则就锁定互斥体并返回值为 true。

4. public func unlock(): Unit

解锁互斥体,如果互斥体被重复加锁了 N 次,那么需要调用 N 次该函数来完全解锁,一旦互斥体被完全解锁,如果有其他线程阻塞在此锁上,则唤醒它们中的一个;如果当前线程没有持有该互斥体,则抛出异常。

在使用互斥锁时,进入临界区前一定要先获取锁(lock 函数),退出临界区时务必释放锁(unlock 函数),获取和释放是一一对应的,如果获取锁后不释放,别的线程就不能再进入临界区。使用互斥锁改写 24.2.1 节中的示例代码,改写后的代码如下:

```
//Chapter24/mutex_demo.cj

from std import collection.ArrayList
```

```
from std import sync.*
from std import time.*

//用来累加计数的累加器
var counter = 0

main() {
    //创建互斥锁
    let countLock = ReentrantMutex()

    //累加器清零
    counter = 0

    //存放线程执行结果的列表
    let resultList = ArrayList<Future<Unit>>()

    //循环 10 次,将创建 10 个线程
    for (i in 0..10) {
        //创建一个用来累加的线程,线程内循环一万次,每次对累加器加 1
        let newThreadFuture = spawn { => for (j in 0..10000) {
            //等待 1μs
            sleep(Duration.microsecond)

            //进入临界区,加锁
            countLock.lock()

            //下面是临界区的代码,本例只有一行累加器加 1 的代码
            //累加器加 1
            counter = counter + 1

            //退出临界区,解锁
            countLock.unlock()
        }}

        //将线程执行结果放入列表
        resultList.append(newThreadFuture)
    }

    //遍历线程执行结果列表,确保每个线程执行完毕
    for (result in resultList) {
        result.get()
    }

    //打印累加器的值
    println(counter)
}
```

编译后运行该示例,输出如下:

```
100000
```

### 24.2.4 监视器

设想在一个典型的生产者/消费者模式中，一类线程负责生产数据，另一类线程负责消费数据，为了简化起见，假设只有一个生产者线程和一个消费者线程，并且生产者线程和消费者线程通过一个变量共享数据，在这种情形下，需要解决两个线程对共享数据的访问引起的数据竞争问题。具体的解决方案有多种，其中一种方案是使消费者线程处于挂起状态，生产者线程生产出数据后，发送通知唤醒消费者线程，消费者线程被唤醒后，消费数据，然后重新进入挂起状态，等待下一次被生产者线程唤醒，这种方案被称为"等待-通知"方案。仓颉语言通过监视器（在操作系统领域也被翻译为管程）的方式，提供了对这种方案的支持，具体实现监视器的类为 Monitor，该类的主要接口如下：

1. public init()

构造函数。

2. public func wait(timeout!: Duration = Duration.Max): Bool

当前线程挂起，直到对应的 notify 函数被调用，或者挂起时间超过 timeout；如果 Monitor 被其他线程唤醒，则返回值为 true；如果超时，则返回值为 false；如果 timeout <= Duration.Zero，则抛出异常；如果当前线程没有持有该互斥体，则抛出异常。

3. public func notify(): Unit

唤醒等待在该 Montior 上的线程，如果当前线程没有持有该互斥体，则抛出异常。

4. public func notifyAll(): Unit

唤醒所有等待在该 Montior 上的线程，如果当前线程没有持有该互斥体，则抛出异常。

Monitor 处理线程同步的原理大体是这样的：可以认为 Monitor 维护了两个线程队列，一个是外部的阻塞队列，另一个是内部的等待队列。当一个线程希望持有 Monitor 的锁时，会被放入外部的阻塞队列，在持有 Monitor 的锁后，会从阻塞队列里移出。当前持有 Monitor 锁的线程，例如叫作 T1，可以执行自己的任务，然后释放锁；如果 T1 线程不满足执行条件，希望等待某个条件满足后再继续执行，则可以调用 Monitor 的 wait 函数，此时 T1 线程会完全释放持有的 Monitor 锁，但是会记录锁的重入次数，然后 T1 线程会进入 Monitor 内部的等待队列，等待唤醒它的信号。此时，Monitor 的锁没有被特定线程持有，可以被别的线程（例如外部阻塞队列里的线程）竞争持有。如果其他持有该 Monitor 锁的线程，例如叫作 T2，调用了 notify 或 notifyAll 函数，向 T1 发出了信号，T1 则会被唤醒，然后竞争该锁，如果竞争成功，就恢复被放入等待队列时的重入次数，然后执行自己的任务；如果竞争失败，T1 则会被放入外部的阻塞队列。

当调用 Monitor 的 wait、notify、notifyAll 函数时，必须确保当前线程已经持有了对应的 Monitor 锁，否则会抛出异常，所以这些函数必须在 Monitor 锁的保护下进行调用。下面给出一个演示示例，在该示例中，主线程会接收用户的输入，并判断是不是整数，如果是整数就放入一个存储该整数的变量，并将新数字的标志设置为 true，然后发送通知唤醒子线程。子线程判断新数字的标志，如果值为 false，就进入挂起等待状态，被唤醒后，会锁定该数字变量,把该变量的数字加入数字列表中,并将新数字标志设置为 false,然后释放锁,随后打印输出数字列表中的内容,最后开始新的循环,子线程重新进入挂起等待状态,示例代码如下:

```
//Chapter24/monitor_demo.cj

from std import collection.ArrayList
from std import sync.*
from std import console.*
from std import convert.*

//输入数字变量,该变量被主线程用来存储用户输入的数字
//子线程把该变量中的数字加入数字列表
var inputNum = 0

//是否有新数字的标志
var hasNewtNum = false

main() {
    //创建监视器
    let monitor = Monitor()

    //存放输入数字的列表
    let numList = ArrayList<Int64>()

    //启动后台,将输入数字添加到列表的线程中
    spawn {
        => while (true) {
            //获取监视器的锁
            monitor.lock()

            //如果有输入数字的标志为 false,则一直等待
            while (!hasNewtNum) {
                //暂时释放锁,挂起线程,等待恢复的信号
                monitor.wait()
            }

            //线程被唤醒,锁恢复,添加用户输入数字到列表
            numList.append(inputNum)

            //处理完输入数字后,将新数字的标志设置为 false
            hasNewtNum = false

            //释放锁
            monitor.unlock()

            //打印列表
            println(numList)
        }
```

```
    }

    //获取用户输入,输入失败则把输入信息置为 q
    var input = (Console.stdIn.readln() ?? "q").trimAscii()

    //如果用户输入 q,则退出
    while (input != "q") {
        //把用户的输入转换为整数
        let optInput = Int64.tryParse(input)

        //处理转换后的数字
        dealWithInputNum(optInput, monitor)

        //重新获取用户输入
        input = (Console.stdIn.readln() ?? "q").trimAscii()
    }
}

//处理转换后的数字
func dealWithInputNum(optInputNum: Option<Int64>, monitor: Monitor) {
    //提取转换后的内容
    match (optInputNum) {
        //如果转换后是整数
        case Some(value) =>
            //获取监视器的锁
            monitor.lock()

            //把提取的值赋给 inputNum
            inputNum = value

            //将新数字的标志设置为 true
            hasNewtNum = true

            //给线程发通知
            monitor.notifyAll()

            //解锁
            monitor.unlock()

        //转换失败
        case None => return
    }
}
```

编译后运行该示例,依次输入 1、2、3、4、5、q,输出如下:

```
1
[1]
2
[1, 2]
```

```
3
[1, 2, 3]
4
[1, 2, 3, 4]
5
[1, 2, 3, 4, 5]
q
```

### 24.2.5　synchronized 关键字

在并发编程的场景下，互斥锁 ReentrantMutex 因其便利灵活的特点得到了充分的应用，但是，ReentrantMutex 的加锁解锁必须成对进行，有可能出现开发者忘记解锁的情况；同样，在持有锁的情况下抛出异常，也可能出现不能自动释放锁的问题。下面通过一个示例，演示某个线程因为抛出异常而不能自动释放锁的情景，代码如下：

```
//Chapter24/mutex_exception_demo.cj

from std import collection.ArrayList
from std import sync.*
from std import time.*

main() {
    //要锁定的字符串列表
    let strList = ["Cangjie", "is", "the", "best", "programming", "language"]

    //存放线程执行结果的列表
    let resultList = ArrayList<Future<Unit>>()

    //字符串列表锁
    let strListLock = ReentrantMutex()

    //循环启动 6 个线程，每个线程打印自己序号对应的字符串
    for (i in 0..=5) {
        let newThreadFuture = spawn {
            =>
            //等待 1ms
            sleep(Duration.millisecond)
            try {
                //进入临界区，加锁
                strListLock.lock()

                //如果是第 3 个线程，就主动触发异常
                //该操作将会导致本线程直接跳到 catch 块而不能执行解锁动作
                if (i == 2) {
                    throw Exception("Thread ${i}:Customer Exception")
                }

                //打印线程序号对应的字符串
```

```
                println(strList[i])

                //退出临界区,解锁
                strListLock.unlock()
            } catch (err: Exception) {
                println(err)
            }
        }

        //将线程执行结果放入列表
        resultList.append(newThreadFuture)
    }

    //遍历线程执行结果列表,确保每个线程执行完毕
    for (result in resultList) {
        result.get()
    }
}
```

编译后运行该示例,可能的输出如下:

```
Cangjie
is
Exception Thread 2:Customer Exception
```

从以上示例可以看出,第 3 个线程因为出现异常,直接跳到了 catch 块,导致没有机会执行退出临界区的解锁代码,这样,第 3 个线程就一直持有锁,其他线程如果在之后再来获取锁,就会一直处于等待状态,这也就是程序执行时会处于"假死"状态的原因。

为解决 ReentrantMutex 加解锁可能出现的问题,仓颉语言提供了 synchronized 关键字,搭配 ReentrantMutex 一起使用,可以在其后跟随的作用域内自动进行加锁解锁操作,使用方式如下:

```
synchronized(mutex){
    //临界区代码
}
```

其中,mutex 是 ReentrantMutex 类的实例,在进入 synchronized 代码块之前,线程会自动获取 mutex 对应的锁,如果无法获取,则该线程会阻塞;在退出 synchronized 代码块之前,线程会自动释放 mutex 对应的锁。下面对本节前面的示例进行改写,通过 synchronized 关键字保证在出现异常时也可以释放获取的锁,示例代码如下:

```
//Chapter24/synchronized_demo.cj

from std import collection.ArrayList
from std import time.*
from std import sync.*

main() {
```

```
        //要锁定的字符串列表
        let strList = ["Cangjie", "is", "the", "best", "programming", "language"]

        //存放线程执行结果的列表
        let resultList = ArrayList<Future<Unit>>()

        //字符串列表锁
        let strListLock = ReentrantMutex()

        //循环启动 6 个线程,每个线程打印自己序号对应的字符串
        for (i in 0..=5) {
            let newThreadFuture = spawn { =>
                sleep(Duration.millisecond)                     //等待 1ms
                try {
                    //使用 synchronized 关键字保证加解锁正常执行
                    synchronized(strListLock) {
                        //如果是第 3 个线程,就主动触发异常
                        //该操作将会导致本线程直接跳到 catch 块而不能执行打印动作
                        if (i == 2) {
                            throw Exception("Thread ${i}:Customer Exception")
                        }

                        //打印线程序号对应的字符串
                        println(strList[i])
                    }
                } catch (err: Exception) {
                    println(err)
                }
            }

            //将线程执行结果放入列表
            resultList.append(newThreadFuture)
        }

        //遍历线程执行结果列表,确保每个线程执行完毕
        for (result in resultList) {
            result.get()
        }
    }
```

编译后运行该示例,可能的输出如下:

```
is
Cangjie
best
programming
language
Exception: Thread 2:Customer Exception
```

# 第 25 章 文件处理

文件处理是大部分软件都涉及的基础功能,是数据存储、读取、修改、维护所必不可少的,本章将讲解基础的目录、文件处理,通过具体示例的方式演示在仓颉语言中相关接口的用法。

本章介绍的文件处理类都属于 fs 库,位于 std 模块的 fs 包下,导入方式如下:

```
from std import fs.*
```

## 25.1 FileInfo

FileInfo 是对应文件系统中文件元数据的结构体,提供文件属性的查询和操作,主要成员如下所示。

1. public init(path: String)
使用字符串形式的目录路径 path 创建 FileInfo 实例,路径非法时抛出异常。

2. public init(path: Path)
使用 Path 形式的目录路径 path 创建 FileInfo 实例,路径非法时抛出异常。

3. public prop parentDirectory: Option\<FileInfo>
父级目录元数据,有父级时返回 Option.Some(v),否则返回 Option.None。

4. public prop path: Path
当前文件或目录路径,以 Path 形式返回

5. public prop symbolicLinkTarget: Option\<Path>
链接目标路径,当前是符号链接返回 Option.Some(v),否则返回 Option.None。

6. public prop creationTime: DateTime
文件或目录的创建时间。

7. public prop lastAccessTime: DateTime
文件或目录的最后访问时间。

8. public prop lastModificationTime: DateTime

文件或目录的最后修改时间。

9. public prop length: Int64

当前是文件时,表示单个文件占用磁盘空间的大小;当前是目录时,表示当前目录的所有文件占用磁盘空间的大小(不包含子目录)。

10. public func isSymbolicLink(): Bool

当前文件是否是软链接。

11. public func isFile(): Bool

当前文件是否是普通文件。

12. public func isDirectory(): Bool

当前文件是否是目录。

13. public func isReadOnly(): Bool

当前文件或目录是否只读,对于 Windows 系统的目录,返回值始终为 false。

14. public func isHidden(): Bool

当前文件或目录是否隐藏。

15. public func canExecute(): Bool

当前用户是否有权限执行该实例对应的文件,对文件而言,判断用户是否有执行文件的权限;对目录而言,判断用户是否有进入目录的权限;在 Windows 环境下,用户对于文件的执行权限由文件扩展名决定,用户始终拥有对于目录的执行权限,该函数不生效,返回值为 true。

16. public func canRead(): Bool

当前用户是否有权限读取该实例对应的文件,对文件而言,判断用户是否有读取文件的权限;对目录而言,判断用户是否有浏览目录的权限;在 Windows 环境下,用户始终拥有对于文件和目录的可读权限,该函数不生效,返回值为 true。

17. public func canWrite(): Bool

当前用户是否有权限写入该实例对应的文件,对文件而言,判断用户是否有写入文件的权限;对目录而言,判断用户是否有删除、移动、创建目录内文件的权限;在 Windows 环境下,用户对于文件的可写权限正常使用,用户始终拥有对于目录的可写权限,该函数不生效,返回值为 true。

18. public func setExecutable(executable: Bool): Bool

对当前实例对应的文件设置当前用户是否可执行的权限,当前用户没有权限修改时抛出异常;对文件而言,设置用户是否有执行文件的权限,对目录而言,设置用户是否有进入目录的权限;在 Windows 环境下,用户对于文件的执行权限由文件扩展名决定,用户始终拥有对目录的执行权限,该函数不生效,返回值为 false;在 Linux 环境下,如果在此函数调用期间,该 FileInfo 对应的文件实体被其他用户或者进程修改,则有可能因为竞争条件(Race Condition)导致其他修改不能生效。

19. public func setReadable(readable: Bool): Bool

对当前实例对应的文件设置当前用户是否有可读取的权限,当前用户没有权限修改时抛出异常;对文件而言,设置用户是否有读取文件的权限;对目录而言,设置用户是否有浏览目录的权限;在 Windows 环境下,用户始终拥有对于文件及目录的可读权限,不可更改,该函数不生效,当参数 readable 为 true 时,函数的返回值为 true,当 readable 为 false 时,函数的返回值为 false;在 Linux 环境下,如果在此函数调用期间,该 FileInfo 对应的文件实体被其他用户或者进程修改,则有可能因为竞争条件导致其他修改不能生效。

20. public func setWritable(writable: Bool): Bool

对当前实例对应的文件设置当前用户是否有可写入的权限,当前用户没有权限修改时抛出异常;对文件而言,设置用户是否有写入文件的权限;对目录而言,设置用户是否有删除、移动、创建目录内文件的权限;在 Windows 环境下,用户对于文件的可写权限正常使用;用户始终拥有对于目录的可写权限,不可更改,该函数不生效,返回值为 false;在 Linux 环境下,如果在此函数调用期间,该 FileInfo 对应的文件实体被其他用户或者进程修改,则有可能因为竞争条件导致其他修改不能生效。

FileInfo 的应用示例代码如下,该示例把文件或者目录作为启动参数,然后打印文件或者目录的信息:

```
//Chapter25/fileinfo_demo.cj

from std import fs.*

main(args: Array<String>) {
    //把文件或者目录作为启动参数
    if (args.size > 0) {
        //实例化文件信息
        let fileInfo = FileInfo(args[0])

        //输出路径
        println("Path: ${fileInfo.path}")

        //是否为文件
        println("Is File: ${fileInfo.isFile()}")

        //最后修改时间
        println("Modification Time: ${fileInfo.lastModificationTime}")

        //是否为只读
        println("ReadOnly: ${fileInfo.isReadOnly()}")

        //是否隐藏文件
        println("Hidden: ${fileInfo.isHidden()}")

        //文件大小
        println("size: ${fileInfo.length}")
```

```
            }
        }
```

编译运行该文件,在 Windows 系统下分别把 C:\Windows\zh-CN\ 和 C:\Windows\zh-CN\notepad.exe.mui 作为启动参数(Linux 下可以选择别的文件和目录),可能的输出如下:

```
.\main.exe C:\Windows\zh-CN
Path:C:\Windows\zh-CN
Is File:false
Modification Time:2022-12-14T23:55:54Z
ReadOnly:false
Hidden:false
size:70656
.\main.exe C:\Windows\zh-CN\notepad.exe.mui
Path:C:\Windows\zh-CN\notepad.exe.mui
Is File:true
Modification Time:2019-12-07T11:53:00Z
ReadOnly:false
Hidden:false
size:8704
```

## 25.2 File

File 是进行文件操作的类,例如文件的创建、删除、移动、流式读写等,创建的实例对象会默认打开对应的文件。使用结束后需要及时调用 close 函数关闭文件,否则会造成资源泄露。主要成员如下所示。

1. public init(path: String, openOption: OpenOption)

使用字符串类型的文件路径 path 及文件打开选项 openOption 创建 File 对象; openOption 是 OpenOption 类型的枚举,包括以下的枚举成员:

- Append:文件尾部追加;
- Create(Bool):创建新文件,参数表示是否具有 Read 权限;
- Truncate(Bool):覆盖文件,参数表示是否具有 Read 权限;
- Open(Bool, Bool):打开现有文件,第 1 个参数指定文件是否具有 Read 权限,第 2 个参数指定文件是否具有 Write 权限;
- CreateOrTruncate(Bool):如果文件不存在,则创建新文件;如果此文件已存在,则将其覆盖,通过参数指定是否具有 Read 权限;
- CreateOrAppend:如果文件不存在,则创建新文件;如果此文件已存在,则在文件尾部追加。

2. public init(path: Path, openOption: OpenOption)

使用 Path 类型的文件路径 path 及文件打开选项 openOption 创建 File 对象。

3. public static func exists(path：String)：Bool

判断字符串类型的文件路径 path 对应的文件是否存在。路径不存在或者路径对应的不是文件返回值为 false，路径对应的是文件或指向文件的软链接返回值为 true。

4. public static func exists(path：Path)：Bool

判断 Path 类型的文件路径 path 对应的文件是否存在。路径不存在或者路径对应的不是文件返回值为 false，路径对应的是文件或指向文件的软链接返回值为 true。

5. public func close()：Unit

关闭当前对象。

6. public func isClosed()：Bool

判断当前对象是否已关闭，未关闭返回值为 false，已关闭返回值为 true。

7. public prop length：Int64

文件当前光标位置至文件尾的数据字节数，一般情况下，文件打开时光标是在文件的开始位置，但是，使用 Append 模式打开时，光标在文件的末尾。下面演示同一个文件在不同打开模式下属性 length 的值，示例代码如下：

```
//Chapter25/file_length_demo.cj

from std import fs.*

main(args: Array<String>) {
    //把文件或者目录作为启动参数
    if (args.size > 0) {
        let filePath = args[0]
        if (!File.exists(filePath)) {
            println("File does not exist!")
            return
        }

        //实例化文件信息，使用追加模式打开
        var file = File(filePath, OpenOption.Append)

        //输出路径
        println("追加模式打开,到文件末尾字节数: ${file.length}")
        file.close()

        //实例化文件信息，使用正常模式打开
        file = File(filePath, OpenOption.Open(false, false))

        //输出路径
        println("正常模式打开,到文件末尾字节数: ${file.length}")
        file.close()
    }
}
```

编译该文件后，在运行前需要先创建一个文件，例如 d:\test.txt，保证该文件有内容

（假设有 10 字节的内容），然后运行编译后的文件，把 d:\test.txt 作为运行参数输入，可能的输出如下：

```
.\main.exe d:\test.txt
追加模式打开,到文件末尾字节数:0
正常模式打开,到文件末尾字节数:10
```

8. public prop info: FileInfo

文件元数据信息。

9. public func read(buffer: Array<Byte>): Int64

从文件中读出数据到 buffer 中，读取成功返回读取字节数，如果文件被读完，则返回 0；如果 buffer 为空，则抛出 IllegalArgumentException 异常；读取失败、文件已关闭，或文件不可读，则抛出 FSException 异常。

10. public func write(buffer: Array<Byte>): Unit

将 buffer 中的数据写入文件中，若 buffer 为空则直接返回；如果写入失败或者只写入了部分数据，或文件已关闭、文件不可写，则抛出 FSException 异常。

11. public func flush(): Unit

清空缓存区，该函数提供默认实现，默认实现为空。

12. public func seek(sp: SeekPosition): Int64

将光标跳转到指定位置 sp 并返回文件头部到跳转后位置的偏移量（以字节为单位），指定的位置不能位于文件头部之前，但可以超过文件末尾；指定位置到文件头部的最大偏移量不能超过当前文件系统允许的最大值，这个最大值接近当前文件系统所允许的最大文件大小，一般为最大文件大小减去 4096 字节。指定位置不满足以上情况，或文件不能跳转时，均会抛 FSException 异常。

SeekPosition 是枚举类型，表示光标在文件中的位置，位于 std 模块的 io 包内，包括以下的枚举成员：

- Current(offset: Int64)：在当前实际位置基础上偏移 offset 的位置；
- Begin(offset: Int64)：在文件开始位置基础上偏移 offset 的位置；
- End(offset: Int64)：在文件结束位置基础上偏移 offset 的位置。

13. public func canRead(): Bool

当前对象是否可读，该函数返回值由创建文件对象的 openOption 所决定，文件对象关闭后返回值为 false。

14. public func canWrite(): Bool

当前对象是否可写，该函数返回值由创建文件对象的 openOption 所决定，文件对象关闭后返回值为 false。

15. public func readToEnd(): Array<Byte>

读取当前 File 所有剩余数据，以 Array 形式返回剩余的 bytes；如果读取失败，或文件不可读，则抛出 FSException 异常。

**16. public func copyTo(out: OutputStream): Unit**

将当前 File 还没读取的数据全部写入指定的输出流 out 中,文件读取时未被打开或文件没有读取权限时,则抛出 FSException 异常。

**17. public static func openRead(path: String): File**

以只读模式打开指定路径 path 代表的文件并返回 File 实例,该实例需要及时调用 close 函数关闭文件。如果文件不存在,或文件不可读,则抛出 FSException 异常;如果文件路径包含空字符,则抛出 IllegalArgumentException 异常。

**18. public static func openRead(path: Path): File**

以只读模式打开指定路径 path 代表的文件并返回 File 实例,该实例需要及时调用 close 函数关闭文件。如果文件不存在,或文件不可读,则抛出 FSException 异常;如果文件路径包含空字符,则抛出 IllegalArgumentException 异常。

**19. public static func create(path: String): File**

以只写模式创建指定路径 path 代表的文件并返回 File 实例。如果路径指向的文件的上级目录不存在或文件已存在,则抛出 FSException 异常;如果文件路径为空字符串或包含空字符,则抛出 IllegalArgumentException 异常。

**20. public static func create(path: Path): File**

以只写模式创建指定路径 path 代表的文件并返回 File 实例。如果路径指向的文件的上级目录不存在或文件已存在,则抛出 FSException 异常;如果文件路径为空字符串或包含空字符,则抛出 IllegalArgumentException 异常。

**21. public static func createTemp(directoryPath: String): File**

在指定目录 directoryPath 下创建临时文件并返回 File 实例,创建的文件名称如 tmpFileXXXXXX 形式,不使用的临时文件应手动删除;创建文件失败或路径不存在则抛出 FSException 异常;如果文件路径为空字符串或包含空字符,则抛出 IllegalArgumentException 异常。

**22. public static func createTemp(directoryPath: Path): File**

在指定目录 directoryPath 下创建临时文件并返回 File 实例,创建的文件名称如 tmpFileXXXXXX 形式,不使用的临时文件应手动删除;创建文件失败或路径不存在则抛出 FSException 异常;如果文件路径为空字符串或包含空字符,则抛出 IllegalArgumentException 异常。

下面通过一个示例演示创建文件和临时文件的方法,示例代码如下:

```
//Chapter25/create_file_demo.cj

from std import fs.*
from std import os.*

main(args: Array<String>) {
    //当前目录(如果目录包含中文,则可能抛出异常)
    let currentPath = currentDir().info.path.toString()
```

```
        //如果包含启动参数,则创建参数指定的文件
        if (args.size > 0) {
            let filePath = currentPath + getPathSeparator() + args[0]
            if (!File.exists(filePath)) {
                let newFile = File.create(filePath)
                println("Created new file: ${newFile.info.path}")
                newFile.close()
            } else {
                println("File already exists!")
            }
        } else { //如果不包含启动参数,则在当前目录创建临时文件
            let newFile = File.createTemp(currentPath)
            let filePath = newFile.info.path
            println("Created temp file: ${filePath}")
            newFile.close()
            File.delete(filePath)
        }
}

//根据不同的编译环境返回对应的路径分隔符,Windows 系统下是"\"
@When[os == "windows"]
func getPathSeparator() {
    return "\\"}

//根据不同的编译环境返回对应的路径分隔符,Linux 系统下是"/"
@When[os != "windows"]
func getPathSeparator() {
    return "/"
}
```

在本示例中,如果包含启动参数,则把启动参数作为文件名,在当前目录里创建该文件;如果不包含启动参数,则在当前目录里创建临时文件。需要注意的是,在 Windows 系统和 Linux 系统中,对于路径分隔符的使用是不同的,Windows 系统中使用"\",而 Linux 系统中使用"/"。所以,对于函数 getPathSeparator 使用了条件编译,根据编译环境的不同返回对应的分隔符,关于条件编译的详细用法会在后续章节介绍。

假如在 Windows 系统下编译该示例,然后运行两次,第 1 次不包含启动参数,第 2 次包含启动参数,可能的输出如下(输出内容根据运行所在目录或者启动参数不同而不同):

```
d:\test> main.exe
Created temp file:D:\test\tmpFilea01292
d:\test> main.exe demo.txt
Created new file:d:\test\demo.txt
```

**23. public static func readFrom(path: String): Array < Byte >**

直接读取指定路径 path 对应的文件,以 Array 形式返回所有的字节。如果文件读取失败、文件关闭失败、文件路径为空、文件不可读,则抛出 FSException 异常;如果文件路径包含空字符,则抛出 IllegalArgumentException 异常。

24. public static func readFrom(path: Path): Array<Byte>

直接读取指定路径 path 对应的文件，以 Array 形式返回所有的字节。如果文件读取失败、文件关闭失败、文件路径为空、文件不可读，则抛出 FSException 异常；如果文件路径包含空字符，则抛出 IllegalArgumentException 异常。

25. public static func writeTo(path: String, buffer: Array<Byte>, openOption!:
    OpenOption = CreateOrAppend): Unit

按照文件打开选项 openOption 打开指定路径 path 代表的文件，并将 buffer 写入该文件。如果文件路径为空，则抛出 FSException 异常；如果文件路径为空字符串或包含空字符，则抛出 IllegalArgumentException 异常。

26. public static func writeTo(path: Path, buffer: Array<Byte>, openOption!:
    OpenOption = CreateOrAppend): Unit

按照文件打开选项 openOption 打开指定路径 path 代表的文件，并将 buffer 写入该文件。如果文件路径为空，则抛出 FSException 异常；如果文件路径为空字符串或包含空字符，则抛出 IllegalArgumentException 异常。

27. public static func delete(path: String): Unit

删除字符串类型的指定路径 path 代表的文件，删除文件前需保证文件已关闭，否则可能删除失败。如果指定文件不存在，或删除失败，则抛出 FSException 异常。

28. public static func delete(path: Path): Unit

删除 Path 类型的指定路径 path 代表的文件，删除文件前需保证文件已关闭，否则可能删除失败。如果指定文件不存在，或删除失败，则抛出 FSException 异常。

29. public static func move(sourcePath: String, destinationPath: String, overwrite:
    Bool): Unit

移动源文件 sourcePath 到新的目标路径 destinationPath。当 overwrite 为 true 时，如果指定路径中已有同名的文件，则会覆盖已存在的文件；如果目标目录名为空，或目标目录名包含空字符，则抛出 IllegalArgumentException 异常；如果源目录不存在，或 overwrite 为 false 时目标目录已存在，或移动失败，则抛出 FSException 异常。

30. public static func move(sourcePath: Path, destinationPath: Path, overwrite:
    Bool): Unit

移动源文件 sourcePath 到新的目标路径 destinationPath。当 overwrite 为 true 时，如果指定路径中已有同名的文件，则会覆盖已存在的文件；如果目标目录名为空，或目标目录名包含空字符，则抛出 IllegalArgumentException 异常；如果源目录不存在，或 overwrite 为 false 时目标目录已存在，或移动失败，则抛出 FSException 异常。

31. public static func copy(sourcePath: String, destinationPath: String, overwrite:
    Bool): Unit

将文件从源路径 sourcePath 复制到新的位置 destinationPath。当 overwrite 为 true 时，如果指定路径中已有同名的文件，则会覆盖已存在的文件；如果文件路径为空，或源路径文件无读权限,或目标路径文件已存在且无写权限,则抛出 FSException 异常。如果文件路径包含空字符,则抛出 IllegalArgumentException 异常。

## 25.3 Directory

Directory 是进行目录操作的类,还提供了遍历目录的能力,主要成员如下所示。

1. **public init(path: String)**

使用字符串类型的目录路径 path 创建 Directory 对象,路径非法时抛出 FSException 异常。

2. **public init(path: Path)**

使用 Path 类型的目录路径 path 创建 Directory 对象,路径非法时抛出 FSException 异常。

3. **public static func exists(path: String): Bool**

判断字符串类型的目录路径 path 对应的目录是否存在。如果路径不存在,或者路径对应的不是目录,则返回值为 false;如果路径对应的是目录或指向目录的软链接,则返回值为 true。

4. **public static func exists(path: Path): Bool**

判断 Path 类型的目录路径 path 对应的目录是否存在。如果路径不存在,或者路径对应的不是目录,则返回值为 false;如果路径对应的是目录或指向目录的软链接,则返回值为 true。

5. **public static func create(path: String, recursive!: Bool = false): Directory**

创建字符串类型的目录路径 path 对应的目录并返回新创建的 Directory 实例,recursive 指定是否递归创建。如果递归创建,将逐级创建路径中不存在的目录,否则仅创建最后一级目录;当目录已存在,或非递归时中间有不存在的目录时,抛出 FSException 异常;当目录为空、目录为当前目录、目录为根目录,或目录中存在空字符时,抛出 IllegalArgumentException 异常。

6. **public static func create(path: Path, recursive!: Bool = false): Directory**

创建 Path 类型的目录路径 path 对应的目录并返回新创建的 Directory 实例,recursive 指定是否递归创建。如果递归创建,将逐级创建路径中不存在的目录,否则仅创建最后一级目录;当目录已存在,或非递归时中间有不存在的目录时,抛出 FSException 异常;当目录为空、目录为当前目录、目录为根目录,或目录中存在空字符时,抛出 IllegalArgumentException 异常。

7. **public static func createTemp(directoryPath: String): Directory**

在字符串类型的目录路径 directoryPath 下创建临时目录并返回临时目录的 Directory 实例。当目录不存在或创建失败时,抛出 FSException 异常;当目录为空或包含空字符时,抛出 IllegalArgumentException 异常。

8. public static func createTemp(directoryPath: Path): Directory

在 Path 类型的目录路径 directoryPath 下创建临时目录并返回临时目录的 Directory 实例。当目录不存在或创建失败时，抛出 FSException 异常；当目录为空或包含空字符时，抛出 IllegalArgumentException 异常。

9. public static func delete(path: String, recursive!: Bool = false): Unit

删除字符串类型的目录路径 path 对应的目录，recursive 指定是否递归删除。如果递归删除，则将逐级删除目录，否则仅删除最后一级目录；如果目录不存在或递归删除失败，则抛出 FSException 异常；如果 path 为空字符串或包含空字符或长度大于 4096，则抛出 IllegalArgumentException 异常。

10. public static func delete(path: Path, recursive!: Bool = false): Unit

删除 Path 类型的目录路径 path 对应的目录，recursive 指定是否递归删除。如果递归删除，则将逐级删除目录，否则仅删除最后一级目录；目录不存在或递归删除失败，抛出 FSException 异常；如果 path 为空字符串或包含空字符或长度大于 4096，则抛出 IllegalArgumentException 异常。

11. public static func move(sourceDirPath: String, destinationDirPath: String, overwrite: Bool): Unit

将字符串类型的源目录路径 sourceDirPath 及文件、子文件夹都移动到指定目标路径 destinationDirPath。overwrite 指定是否覆盖，为 true 时将覆盖目标路径中的所有子文件夹和文件；如果目标目录名为空，或目标目录名包含空字符，则抛出 IllegalArgumentException 异常；如果源目录不存在，或 overwrite 为 false 时目标目录已存在，或移动失败，则抛出 FSException 异常。

12. public static func move(sourceDirPath: Path, destinationDirPath: Path, overwrite: Bool): Unit

将 Path 类型的源目录路径 sourceDirPath 及文件、子文件夹都移动到指定目标路径 destinationDirPath。overwrite 指定是否覆盖，为 true 时将覆盖目标路径中的所有子文件夹和文件；如果目标目录名为空，或目标目录名包含空字符，则抛出 IllegalArgumentException 异常；如果源目录不存在，或 overwrite 为 false 时目标目录已存在，或移动失败，则抛出 FSException 异常。

13. public static func copy(sourceDirPath: String, destinationDirPath: String, overwrite: Bool): Unit

将字符串类型的源目录路径 sourceDirPath 及文件、子文件夹都复制到目标位置 destinationDirPath。overwrite 指定是否覆盖，为 true 时将覆盖目标路径中的所有子文件夹和文件；如果源目录不存在，或 overwrite 为 false 时目标目录已存在，或目标目录在源目录中，或复制失败，则抛出 FSException 异常；如果源目录或目标目录名包含空字符，则抛出 IllegalArgumentException 异常。

14. public static func copy(sourceDirPath：Path，destinationDirPath：Path，overwrite：Bool)：Unit

将 Path 类型的源目录路径 sourceDirPath 及文件、子文件夹都复制到目标位置 destinationDirPath。overwrite 指定是否覆盖，为 true 时将覆盖目标路径中的所有子文件夹和文件；如果源目录不存在，或 overwrite 为 false 时目标目录已存在，或目标目录在源目录中，或复制失败，则抛出 FSException 异常；如果源目录或目标目录名包含空字符，则抛出 IllegalArgumentException 异常。

15. public func createSubDirectory(name：String)：Directory

在当前目录下创建 name 代表的子目录并返回子目录的 Directory 实例，name 只接收不带路径前缀的字符串。当子目录名中含有路径信息，路径已存在，或创建目录失败时，抛出 FSException 异常；当目录名中含有空字符时，抛出 IllegalArgumentException 异常。

16. public func createFile(name：String)：File

在当前目录下创建 name 代表的子文件并返回子文件的 File 实例，该实例需要手动调用 close 函数关闭文件，name 只接收不带路径前缀的字符串。当子文件名中含有路径信息，或文件名已存在时，抛出 FSException 异常；当文件名包含空字符时，抛出 IllegalArgumentException 异常。

17. public prop info：FileInfo

当前目录的元数据信息。

18. public func isEmpty()：Bool

当前目录是否为空。

19. public func iterator()：Iterator＜FileInfo＞

当前目录的文件或子目录迭代器。

20. public func directories()：Iterator＜FileInfo＞

当前目录的子目录迭代器。

21. public func files()：Iterator＜FileInfo＞

当前目录的子文件迭代器。

22. public func entryList()：ArrayList＜FileInfo＞

当前目录的文件或子目录列表。

23. public func directoryList()：ArrayList＜FileInfo＞

当前目录的子目录列表。

24. public func fileList()：ArrayList＜FileInfo＞

当前目录的子文件列表。

下面通过一个示例演示 Directory 函数的常用用法，示例代码如下：

```
//Chapter25/directory_demo.cj

from std import fs.*
```

```
from std import os.*

main(): Unit {
    //要演示目录管理的目录全路径名
    let dirName = "/data/code/demo/"

    //如果目录不存在,则创建该目录
    if (!Directory.exists(dirName)) {
        //创建目录
        try {
            Directory.create(dirName, recursive: true)
            println("Created successfully")
        } catch (err: Exception) {
            //创建目录失败,退出程序
            println("Creation failed ${err}")
            return
        }
    }

    //实例化目录对象
    let dirInfo = Directory(dirName)

    //获取子目录并打印输出
    let subDirInfoList = dirInfo.directories()
    for (subDirInfo in subDirInfoList) {
        println(subDirInfo.path)
    }

    //获取目录内文件并打印输出
    let subFileInfoList = dirInfo.files()
    for (subFileInfo in subFileInfoList) {
        println(subFileInfo.path)
    }

    //获取 subdir 子目录
    let subDir = dirName + "subdir"
    if (!Directory.exists(subDir)) {
        try {
            Directory.create(subDir, recursive: true)
            println("subDir Created successfully")
        } catch (err: Exception) {
            //创建目录失败,退出程序
            println("Creation failed ${err}")
            return
        }
    }

    println("The sub directory is ${subDir}")
```

```
    //子目录要移动到的位置
    let newDestDirName = "/data/code/testdir/"

    //移动子目录到新的位置
    Directory.move(subDir, newDestDirName, true)
    println("The destination directory is ${newDestDirName}")

    //删除移动到的新目录
    try {
        Directory.delete(newDestDirName, recursive: true)
        println("The destination folder was successfully deleted")
    } catch (err: Exception) {
        //删除目录失败,退出程序
        println("deleted failed ${err}")
        return
    }
}
```

假设该实例使用的目录运行在 Linux 系统下,如果要运行在 Windows 系统下,可以根据实际情况修改。在 Linux 系统下编译运行该文件,可能的输出如下:

```
/data/code/demo/src
/data/code/demo/main
/data/code/demo/demo.cj
subDir Created successfully
The sub directory is /data/code/demo/subdir
The destination directory is /data/code/testdir/
The destination folder was successfully deleted
```

## 25.4 文件读写示例

日志记录是软件开发中常用的基础功能,本节将利用本章讲解的文件处理和文件读写的常用函数,开发一个简单的日志处理类,并演示该类的使用方法。在本示例中,日志处理类为 LogTools,它通过单例模式保证日志文件在当前目录的 logs 子目录下创建,示例代码如下:

```
//Chapter25/log_file_demo.cj

from std import time.*
from std import random.Random
from std import fs.*
from std import os.*
from std import convert.*

main(): Unit {
    //记录日志
    LogTools.Instance().logInfo("启动程序")
```

```
    //生成5个随机数并记录到日志中
    let rand = Random()
    for (i in 0..5) {
        let randnum = rand.nextInt32()
        LogTools.Instance().logInfo(randnum.toString())
    }

    //模拟产生异常记录错误日志
    try {
        let divisor = Int64.parse("0")
        let temp = 100 / divisor
    } catch (err: Exception) {
        LogTools.Instance().logError(err.message)
    }

    //输出记录的日志内容
    println(LogTools.Instance().getLogContent())
}

//日志处理类
class LogTools {
    //日志处理类的单例
    private static var instance: Option<LogTools> = Option<LogTools>.None

    //处理日志文件的名称
    private let logFileName: String

    //获取日志处理类的单例对象
    static func Instance(): LogTools {
        match (instance) {
            case Some(tools) => return tools
            case None =>
                let logTools = LogTools()
                instance = Some(logTools)
                return logTools
        }
    }

    init() {
        //获取日志目录,为工作目录下的logs目录
        let logDir = currentDir().info.path.toString() + getPathSeparator() + "logs"

        //如果日志目录不存在,则创建一个
        if (!Directory.exists(logDir)) {
            Directory.create(logDir)
        }

        //日志文件路径
```

```
        logFileName = logDir + getPathSeparator() + "demo.log"
    }

    //记录 info 类型的日志
    func logInfo(info: String) {
        logContent(info, "INFO")
    }

    //记录 error 类型的日志
    func logError(err: String) {
        logContent(err, "ERROR")
    }

    //记录 warning 类型的日志
    func logWarning(warning: String) {
        logContent(warning, "WARNING")
    }

    //实际处理日志文件写入的函数
    private func logContent(content: String, level: String) {
        //以创建或者追加模式打开日志文件,如果文件不存在,则创建该文件并以只写模式打开,
        //否则以追加模式打开
        let logFile = File(logFileName, OpenOption.CreateOrAppend)

        //获取当前时间的字符串形式
        let strTime = DateTime.now().toString("yyyy-MM-dd HH:mm:ss")

        //将日志写到文件
        logFile.write("[ ${strTime}] [ ${level}] ${content}\n".toUtf8Array())

        //清空缓冲区,也就是保证缓冲区数据写入文件
        //当前的实现并没有使用缓存,所以本函数是否调用不会对运行造成影响
        logFile.flush()

        //关闭文件流
        logFile.close()
    }

    //获取日志文件记录的内容
    func getLogContent(): String {
        if (File.exists(logFileName)) {
            //文本形式读取所有内容
            return String.fromUtf8(File.readFrom(logFileName))
        } else {
            return String.empty
        }
    }
}
```

```
//根据不同的编译环境返回对应的路径分隔符,Windows 系统下是"\"
@When[os == "windows"]
func getPathSeparator() {
    return "\\"
}

//根据不同的编译环境返回对应的路径分隔符,Linux 系统下是"/"
@When[os != "windows"]
func getPathSeparator() {
    return "/"
}
```

编译运行该文件,可能的输出如下:

```
[2023-09-16 02:05:02] [INFO] 启动程序
[2023-09-16 02:05:02] [INFO] -1153042320
[2023-09-16 02:05:02] [INFO] 1520831668
[2023-09-16 02:05:02] [INFO] 772209776
[2023-09-16 02:05:02] [INFO] 1298376335
[2023-09-16 02:05:02] [INFO] 1427681778
[2023-09-16 02:05:02] [ERROR] Divided by zero!
```

# 第 26 章 仓颉编译器

cjc 是仓颉自带的编译器,可以把仓颉源程序编译成库文件或者可执行文件,在本书第 3 章对该编译器的用法做了初步的介绍,本章将进一步详细讲解 cjc 的编译选项和常用用法。

## 26.1 编译演示代码

为了方便后续章节演示编译器的各种用法,本节提供了示例代码,代码本身比较简单,包括两个文件,分别是 cjc_demo.cj 和 printer.cj,这两个文件的内容如下:

```
//Chapter26/cjc_demo.cj

import log.printLog

from std import random.Random

main() {
    //实例化随机数发生器
    let rand = Random()

    //得到一个随机数
    let num = rand.nextInt64()

    //打印随机数日志
    printLog(num.toString())
}

//Chapter26/printer.cj

package log

from std import time.*
```

```
//把传入的日志参数按照特定的格式输出
public func printLog(log: String) {
    //获取当前时间
    let now = DateTime.now()

    //把当前时间的转换为格式化字符串
    let logTime = now.toString("yyyy-MM-dd HH:mm:ss")

    //打印输出包含当前时间的日志内容
    println("日志时间:${logTime} 日志内容:${log}")
}
```

两个文件的目录结构如下:

```
demo
 └── src
      ├── cjc_demo.cj
      └── log
           └── printer.cj
```

其中,demo 为项目主目录,读者可以根据实际情况更改为其他的名字,demo 目录下是 src 目录,src 目录下有一个 log 目录和一个源代码文件 cjc_demo.cj,log 目录下是另一个源代码文件 printer.cj。

## 26.2 编译选项

14min

**1. --output-type**

指定输出文件的类型,可以选择的类型如下。
- exe:可执行文件,该类型为缺省类型。
- staticlib:静态链接库(.a 文件)。
- dylib:动态链接库(Linux 系统下为 .so 文件、Windows 系统下为 .dll 文件)。

输出可执行文件是输出文件类型的缺省选择,前面章节已有多个示例,这里以 26.1 节的 printer.cj 为例,演示如何编译成静态链接库和动态链接库。

进入 src 目录,生成静态链接库的命令如下(以 Windows 系统为例,本节默认都是在该系统下编译):

```
cjc .\log\printer.cj --output-type=staticlib
```

编译成功后会在当前目录生成 3 个文件:
- liblog.a
- log.cjo
- log.bchir

其中,liblog.a 即为生成的静态链接库文件。

如果要生成动态链接库,命令如下:

```
cjc .\log\printer.cj --output-type=dylib
```

编译成功后会在当前目录生成 3 个文件:

- liblog.dll
- log.cjo
- log.bchir

其中,liblog.dll 即为生成的动态链接库文件。

在编译可执行文件时,静态链接库可以直接和源文件一起编译。假如已经编译好了静态链接库 liblog.a,则在 src 目录下,该链接库和 cjc_demo.cj 一起编译生成可执行文件的命令如下:

```
cjc -o demo.exe .\cjc_demo.cj .\liblog.a
```

该命令会编译生成文件 demo.exe,运行该可执行文件,命令和输出如下:

```
.\demo.exe
日志时间:2023-09-16 07:07:22 日志内容:-6599183051972175291
```

2. --package,-p

编译一个包,该选项需要一个目录作为参数,并且该目录下所有的源文件属于同一个包。以 26.1 节代码中的 log 目录为例,在 src 目录下,包编译并且生成静态链接库的命令如下:

```
cjc -p .\log\ --output-type=staticlib
```

编译成功后会在当前目录生成 3 个文件:

- liblog.a
- log.cjo
- log.bchir

其中,liblog.a 即为生成的静态链接库文件,输出文件名称的默认格式是(lib+包名+.a)。

3. --library <arg>,-l <arg>,-l <arg>

指定编译时要链接的库文件,其中参数 arg 为库名称,搜索库文件时,在 Linux 系统下会搜索名称为 libarg.a 和 libarg.so 的文件;在 Windows 系统下会搜索 libarg.dll.a、libarg.a、libarg.lib、libarg.dll 及 arg.dll.a、arg.lib、arg.dll 等文件。以 log 库为例,在 Linux 系统下会搜索 liblog.a 的静态链接库文件和 liblog.so 的动态链接库文件。

在搜索库文件时,需要指定搜索的目录,所以该选项很少单独使用,一般会配合 --library-path <arg> 选项使用。除了 --library-path 选项指定的目录外,还会搜索环境变量 LIBRARY_PATH 中指定的路径。

4. --library-path <arg>,-L <arg>,-L <arg>

指定要链接的库文件所在的目录,该选项一般和 --library <arg> 选项配合使用。以

26.1节为例,假如已经编译好了名为liblog.a的静态链接库,并且当前目录是src目录,使用指定链接库文件进行编译的命令如下:

```
cjc .\cjc_demo.cj -o demo.exe -l log -L.
```

该命令会生成demo.exe可执行文件,该文件在笔者计算机上的大小为1 607 972,假如删除了静态链接库liblog.a,然后执行demo.exe,命令及回显如下:

```
rm .\log.cjo
.\demo.exe
日志时间:2023-09-16 09:06:27 日志内容:4688975028923496016
```

可以看到,应用是能够正常执行的。

接下来,重新生成动态链接库文件liblog.dll,然后执行同样的编译命令:

```
cjc -p .\log\ --output-type=dylib
cjc .\cjc_demo.cj -o demo.exe -llog -L.
```

该命令同样会生成demo.exe可执行文件,但是,该文件的大小变为806 242,这时,先运行demo.exe,查看输出:

```
.\demo.exe
日志时间:2023-09-16 09:11:49 日志内容:-352297933242757119
```

应用是可以正常运行的,然后删除liblog.dll动态链接库文件,再运行demo.exe:

```
#rm .\liblog.dll
#.\demo.exe
```

系统会弹出找不到liblog.dll的错误提示窗口,如图26-1所示。

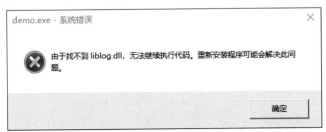

图26-1 找不到liblog.dll

也就是说,在编译仓颉程序时,对于静态链接库会直接编译到输出文件中,这时输出文件体积较大;对于动态链接库,并不会编译到输出文件中,输出文件体积较小,但是运行时要求环境中存在并可成功加载该库文件。

编译时,如果搜索路径同时存在静态链接库和动态链接库,则会优先选择静态链接库进行编译。

**5. --conditional-compilation-config**

用户自定义的条件编译选项,指定编译时的编译条件,详细用法见26.3节"条件编译"。

### 6. -O\<N\>

使用参数 N 指定的优化级别进行编译优化,优化级别越高,编译器越会更多地进行代码优化,生成的程序执行效率越高,编译时间也越长。

当前编译器默认情况下执行 O0 级别的优化,最高支持 O2 级别的优化。

### 7. -O

使用 O0 级别的代码优化,等价于-O0。

### 8. -g

生成带有调试信息的可执行文件或者库文件,该选项只能配合-O0 使用,如果使用更高的优化级别,则可能会导致调试功能出现异常。

### 9. --verbose,-V

该选项会在编译时打印出编译器版本信息,工具链依赖的相关信息及编译过程中执行的命令。该选项不能单独使用,需要在执行编译时使用,示例如下:

```
#cjc -o demo.exe cjc_demo.cj liblog.a -V
Cangjie Compiler: 0.39.7 (4aeaaf53492f 2023-08-24)
"d:\\Program Files (x86)\\Cangjie\\bin\\cjc-frontend.exe" cjc_demo.cj liblog.a -V -o
C:\Users\wanwan\AppData\Local\Temp\cangjie-tmp-d0cb3b88\cjc_demo.bc
info: selected gcc lib path: d:\Program Files
(x86)\Cangjie\third_party\mingw\lib\
""d:\Program Files (x86)\Cangjie\third_party\llvm\bin\opt.exe"
"C:\Users\wanwan\AppData\Local\Temp\cangjie-tmp-d0cb3b88\cjc_demo.bc"
"--cangjie-pipeline" "-passes=default<O0>" "--mtriple=x86_64-windows-gnu" "-o"
"C:\Users\wanwan\AppData\Local\Temp\cangjie-tmp-d0cb3b88\cjc_demo.opt.bc" "
""d:\Program Files (x86)\Cangjie\third_party\llvm\bin\llc.exe"
"C:\Users\wanwan\AppData\Local\Temp\cangjie-tmp-d0cb3b88\cjc_demo.opt.bc"
"--cangjie-pipeline" "-O0" "-disable-Debug-info-print"
"--relocation-model=pic" "--frame-pointer=all" "--filetype=obj"
"--mtriple=x86_64-windows-gnu" "-o"
"C:\Users\wanwan\AppData\Local\Temp\cangjie-tmp-d0cb3b88\cjc_demo.o" "
""d:\Program Files (x86)\Cangjie\third_party\mingw\bin\ld.exe" "-o"
"D:\git\cangjie_practice\\code\Chapter26\demo.exe" "--nxcompat"
"--dynamicbase" "-m" "i386pep" "-Bdynamic" "d:\Program Files
(x86)\Cangjie\third_party\mingw\lib\crt2.o" "d:\Program Files
(x86)\Cangjie\third_party\mingw\lib\crtbegin.o" "-Ld:\Program Files
(x86)\Cangjie\third_party\mingw\lib\" "-Ld:\Program
Files (x86)\Cangjie\lib\windows_x86_64_llvm" "-Ld:\Program Files
(x86)\Cangjie\runtime\lib\windows_x86_64_llvm" "-LD:\mingw64\lib"
"-LD:\mingw64\x86_64-w64-mingw32\lib" "-Ld:\Program Files
(x86)\Cangjie\third_party\mingw\lib\" "-T" "d:\Program Files
(x86)\Cangjie\lib\windows_x86_64_llvm\cjld.lds" "d:\Program Files
(x86)\Cangjie\lib\windows_x86_64_llvm\cjstart.o"
"C:\Users\wanwan\AppData\Local\Temp\cangjie-tmp-d0cb3b88\cjc_demo.o"
"D:\git\cangjie_practice\\code\Chapter26\liblog.a" "--start-group"
"-l:libcangjie-std-core.a" "-l:libcangjie-std-random.a"
"-l:libcangjie-std-core.a" "-l:libcangjie-std-math.a"
```

```
"-l:libcangjie-std-convert.a" "-l:libcangjie-std-unicode.a"
"-l:libcangjie-std-overflow.a" "-l:libcangjie-std-sort.a"
"-l:libcangjie-std-collection.a" "-l:libcangjie-std-time.a" "--end-group"
"-lcangjie-runtime" "-lboundscheck" "-lm" "-lssp_nonshared" "-lssp"
"-lmsvcrt" "-lmingw32" "-lgcc" "-lgcc_eh" "-lmoldname" "-lmingwex"
"-lmingw32" "-lmsvcrt" "-lpthread" "-ladvapi32" "-lshell32" "-luser32"
"-lkernel32" "-lmingw32" "-lgcc" "-lgcc_eh" "-lmoldname" "-lmingwex"
"-lmsvcrt" "-lclang_rt-builtins" "d:\Program Files
(x86)\Cangjie\third_party\mingw\lib\crtend.o" "
```

### 10. -coverage

生成支持统计代码覆盖率的可执行程序。该选项在生成可执行文件时会为每个编译单元生成后缀名为 gcno 的代码信息文件。在运行程序后，每个编译单元会得到一个后缀名为 gcda 的执行统计文件。仓颉语言覆盖率统计工具 cjcov（Cangjie Coverage）利用这两个文件，可以生成本次执行下的代码覆盖率报表。

针对 cj_demo.cj 文件，如果已经生成了 liblog.a 静态链接库，要生成支持统计代码覆盖率的可执行程序，则可以使用的命令如下：

```
cjc -o demo.exe --coverage cjc_demo.cj liblog.a
```

执行该命令后会在当前目录生成两个文件：
- default.gcno
- demo

其中，default.gcno 为代码信息文件。

### 11. --int-overflow=[throwing|wrapping|saturating]

指定固定精度整数运算的溢出策略，有以下 3 种策略可以选择。

（1）throwing 整数运算溢出时会抛出异常，这是默认策略。

（2）wrapping 整数运算溢出时会回转至对应固定精度整数的另外一端。

（3）saturating 整数运算溢出时会选择对应固定精度的极值作为结果。

### 12. -test

对于通常的应用程序，程序执行的入口是 main 函数，但是，如果要对应用程序进行单元测试，就不适合再从原有的 main 函数开始执行了，那样无法执行测试用例。仓颉编译器支持 unittest 测试框架，可以通过宏自动地在编译时生成一个专门用来运行单元测试的 main 函数，该函数会自动执行测试用例，并输出测试统计信息。

下面通过一个简单的示例演示具体的用法，代码如下：

```
//Chapter26/test_demo.cj

from std import unittest.*
from std import unittest.testmacro.*

@Test
public class TestDemo {
```

```
    @TestCase
    public func testCase(): Unit {
        print("test case\n")
    }
}
```

这段代码很简单,只有一个测试用例,也没有 main 函数,要编译成可以进行单元测试的可执行文件,使用的命令如下:

```
#cjc --test test_demo.cj
```

编译成功后,会在当前目录生成可执行文件 main.exe,运行该文件,命令及回显如下:

```
.\main.exe
test case
--------------------------------------------------------------
TP: default, time elapsed: 505600 ns, Result:
    TCS: TestDemo, time elapsed: 489900 ns, RESULT:
    [ PASSED ] CASE: testCase (483500 ns)
Summary: TOTAL: 1
    PASSED: 1, SKIPPED: 0, ERROR: 0
    FAILED: 0
--------------------------------------------------------------
```

这样就打印出了单元测试的结果。

13. --save-temps=<arg>

保留编译过程中的中间文件并存放在参数 arg 指定的目录中,该目录必须已经创建,可以是相对路径也可以是绝对路径。针对 cj_demo.cj 文件,如果已经生成了 liblog.a 静态链接库,要将中间文件保留到 tmp 目录,则可以使用的命令如下:

```
cjc -o demo.exe --save-temps=tmp cjc_demo.cj liblog.a
```

该命令会在当前目录的 tmp 子目录下生成如下 4 个中间文件:

- cjc_demo.bc
- cjc_demo.o
- cjc_demo.opt.bc
- cjc_demo.s

14. --import-path<arg>

在指定了导入模块以后,还需要对代码文件中使用了导入模块的元素进行语义检查与编译,这时就需要导入模块的抽象语法树(Abstract Syntax Tree,AST)文件。默认情况下,编译器从环境中(如 CANGJIE_PATH 环境变量)查找该文件,也可以通过 --import-path 参数直接指定,该参数比环境变量具有更高的优先级。

15. --error-count-limit=<arg>

设置编译器打印错误信息个数的上限。默认情况下编译器最多打印 8 个错误,如果 arg 为 all,则打印所有错误;如果 arg 为非负整数 N,则最多打印 N 个错误。

16. --output-dir < arg >

设置编译器生成的中间文件和最终文件的保存目录。因为--output 选项也可以包含目录,当两者同时存在时,--output 中的目录只能是相对目录。假如--output-dir 选项指定的目录为/data,--output 指定的文件为 tmp\demo,那么命令如下：

```
cjc --output-dir .\data -o tmp\demo.exe cjc_demo.cj liblog.a
```

将会在当前目录的./data/tmp/下生成 demo.exe 文件。

## 26.3 条件编译

因为现代操作系统和架构的复杂性,在企业级开发时,经常会遇到适配多种系统和平台的需求,需要在不同的平台上都可以正常运行。要解决这个问题,一种办法是针对不同的平台提供不同的源文件,在编译时只编译对应的文件即可；另一种办法是在源码里包括所有可能用到的平台代码,在那些需要跨平台适配的代码上,使用宏来标注编译的条件,这样,在编译时指定适配的平台,就可以生成特定平台的输出文件。两种方式各有优缺点,相对来讲,第 2 种方式更有利于代码的统一,维护也比较方便,是目前采用的主要方式。仓颉编译器也针对这种需求,提供了条件编译的支持,目前支持除 package 声明外的顶层条件编译。

### 26.3.1 使用方式

下面通过一个具体的示例来演示条件编译的用法,代码如下：

```
//Chapter26/condition_compile.cj

//操作系统是 Linux 时编译该函数
@when[os == "linux"]
func print() {
  println("Linux")
}

//操作系统是 Windows 时编译该函数
@when[os == "windows"]
func print() {
  println("Windows")
}

main() {
  print()
}
```

编译后运行的命令和回显如下：

```
# cjc condition_compile.cj
# .\main.exe
windows
```

在上述代码里,条件编译的宏是@when,os 是内置的编译条件,代表编译的操作系统,@when[os=="linux"]表示只在操作系统是 Linux 的情况下编译该函数;同样,@when[os=="windows"]表示只在操作系统是 Windows 的情况下编译该函数。@when 支持对顶层的多个节点进行修饰,这时可以使用大括号{}来包裹多个节点,等价于在每个被包裹的节点上加@when 的宏修饰,示例代码如下:

```
@when[os == "linux"]
{
  func print() {
    println("Linux")
  }

  class demo{}
}
```

需要注意的是,@when 宏不支持编译条件的嵌套,如下写法都是错误的:

```
@when[os == "linux"]
@when[debug]
func print() {
    println("Linux")
}

@when[os == "linux"]
{
  @when[debug]
  func print() {
    println("Linux")
  }
}
```

## 26.3.2 内置编译条件

仓颉编译器提供了一些内置的编译条件,可以被直接使用。

### 1. os

操作系统,支持的操作符为==和!=,支持的系统有 windows、linux、macosx、wasi、ios、android、unix、freebsd、hm。

### 2. backend

使用的编译器后端,支持的操作符为==和!=。仓颉语言支持的编译器后端有 llvm、llvm-x86、llvm-x86_64、llvm-arm、llvm-aarch64、cjvm、cjvm-x86、cjvm-x86_64、cjvm-arm、cjvm-aarch64。

### 3. cjc_version

仓颉编译器的版本,可以根据当前仓颉编译器的版本选择要编译的代码,支持的操作符有==、!=、>、<、>=、<=共 6 种。版本号的格式为 xx.xx.xx,每段的两个 x 支持 1~2 位

数字,计算时按照补齐两位计算,补齐的规则是前面加上 0,例如,版本号 0.28.4 和 0.28.04 是相等的,示例代码如下:

```
@when[cjc_version > "0.28.4"]
func print() {
    println("Linux")
}
```

### 4. debug

debug 只支持一元运算符,根据当前编译器是不是 Debug 版本会有不同的输出结果,示例代码如下:

```
@when[debug]
func print() {
    println("Debug")
}

@when[!debug]
func print() {
    println("not Debug")
}
```

## 26.3.3　自定义编译条件

仓颉编译器支持开发者自定义编译条件,使用方式和内置编译条件一样,只是在执行编译命令时需要通过编译器的-config 选项传递自定义编译条件的值,传递的值是 key-value 结构,使用小括号"()"把选项值括起来,如果有多个选项,则使用逗号","分隔。假如自定义了名称为 db 的编译选项,它的两个选项值是 mysql 和 oracle,示例代码如下:

```
//Chapter26/user_defined_condition_compile.cj

//针对编译选项值是 mysql 时选择的编译代码
@when[db == "mysql"]
func execBackup() {
    println("mysql")
}

//针对编译选项值是 oracle 时选择的编译代码
@when[db == "oracle"]
func execBackup() {
    println("oracle")
}

main() {
  execBackup()
}
```

使用自定义编译条件进行编译,分别使用 mysql 和 oracle 的编译选项值进行编译,命令如下：

```
cjc user_defined_condition_compile.cj
--conditional-compilation-config="(db=mysql)" -o mysql_demo.exe

cjc user_defined_condition_compile.cj
--conditional-compilation-config="(db=oracle)" -o oracle_demo.exe
```

这样分别生成了 mysql_demo.exe 和 oracle_demo.exe 文件，运行这两个文件，命令及输出如下：

```
#.\mysql_demo.exe
mysql
#.\oracle_demo.exe
oracle
```

### 26.3.4　多条件编译

仓颉编译器支持对多个编译条件使用逻辑操作符进行自由组合，可以使用括号明确组合的优先级。下面通过一个包括内置编译条件和自定义编译条件的组合，演示多条件编译的用法，示例代码如下：

```
//Chapter26/multi_condition_compile.cj

//针对编译选项 db 值是 mysql 并且操作系统是 Linux 时选择的编译代码
@when[db == "mysql" && os == "linux"]
func execBackup() {
    println("mysql in Linux")
}

//针对编译选项 db 值是 oracle 并且操作系统是 Windows 时选择的编译代码
@when[db == "oracle" && os == "windows"]
func execBackup() {
    println("oracle in Windows")
}

main() {
  execBackup()
}
```

条件编译并运行的命令和回显如下：

```
#cjc multi_condition_compile.cj
--conditional-compilation-config="(db=oracle)" -o multi_demo.exe
#./multi_demo.exe
oracle in windows
```

# 第 27 章 仓颉调试器

程序在实际运行过程中，可能会出现各种与预期不一致的情况，具体的原因比较复杂，一些是程序逻辑设计的问题，另一些可能是无法预见的特殊运行场景，甚至拼写错误也是一种普遍的原因，为了排除运行中的异常，仔细检查代码是一种方法，更常见的是对代码进行调试。通过调试观察程序运行时的状态，检查函数的运行结果、查看实际变量值，从而发现问题，解决问题。

要对应用程序进行调试，常见的思路有两种，一种是调试器复制被调试程序的代码，模拟被调试程序的执行，因为执行过程是模拟的，调试器可以随时查看或者修改被调试程序的运行堆栈和数据信息，这种模式的典型应用是开源运行时诊断工具 Valgrind；另一种是使用操作系统的 ptrace 系统调用，允许调试器监听并控制被调试进程的内存和寄存器，典型应用是开源调试器 LLDB。这两种模式各有优缺点，第 1 种方式的优点是不用重新编译，缺点是运行太慢，比直接运行可能会慢上几十倍；第 2 种方式需要对程序进行编译，并且编译时需要附带调试信息，导致编译后的文件体积较大，优点是运行速度较快，比不带调试信息时运行稍慢一点，两者处于同一数量级。

仓颉提供了仓颉调试器（cjdb），支持对仓颉程序进行调试，仓颉调试器基于开源的 LLDB 实现，目前仅支持 Linux 平台，本章将详细讲解仓颉调试器的常用用法。

## 27.1 仓颉调试器演示代码

为了方便后续演示调试器的各种用法，本章提供了演示示例代码，其功能是对随机数进行累加，输出累加后的值，代码如下：

```
//Chapter27/Debug_demo.cj

from std import random.Random

//全局变量,最大计算次数
let MAX_COUNT = 100
```

```
main() {
    //当前计算次数
    var index = 0

    //累加值
    var sumNum = 0

    //计算 MAX_COUNT 个随机值的累加值
    while (index < MAX_COUNT) {
        sumNum = addRandomNum(sumNum)
        index++
    }
    //打印累加值
    println(sumNum)
}

//产生一个随机数并与参数 currentSum 相加,然后返回两者的和
func addRandomNum(currentSum: Int64): Int64 {
    let rand = Random()
    let newSum = currentSum + rand.nextInt64(100)
    return newSum
}
```

## 27.2 调试版本的编译

仓颉调试器调试程序需要使用被调试程序中的调试信息,所以编译 Debug 版本仓颉程序时需要使用-g 参数,这样,编译时就会把调试信息附加到编译后的程序里。针对本示例,编译命令如下:

```
cjc -g Debug_demo.cj -o demo
```

在演示机环境中,编译后的大小为 631 568 字节,如果不使用-g 参数,直接编译程序,则大小为 629 296 字节,如图 27-1 所示。

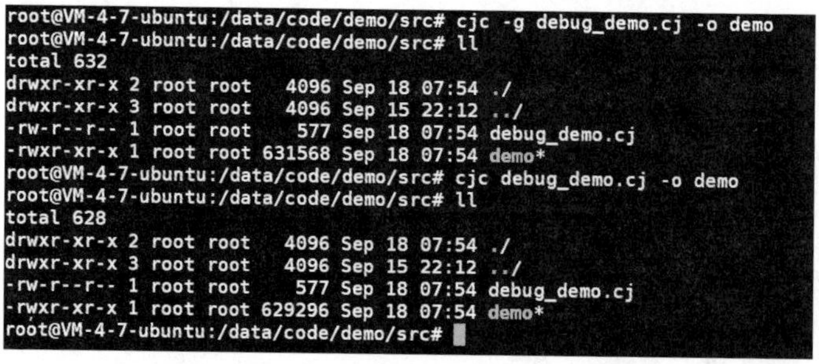

图 27-1 调试信息对文件大小的影响

由此可以看出,Debug 版本的程序附带了大量的调试信息。

## 27.3 启动调试的方式

对于尚未启动的程序及已经启动的程序,仓颉调试器都支持调试,具体可以分为以下 3 种情形。

(1) 对于尚未启动的程序,可以在启动调试器的同时拉起被调试程序,此时被调试程序作为参数传递给调试器,命令及回显如下:

```
root@VM-4-7-Ubuntu:/data/code/demo/src#cjdb demo
(cjdb) target create "demo"
Current executable set to '/data/code/demo/src/demo' (x86_64).
(cjdb)
```

(2) 对于尚未启动的程序,可以在启动调试器后,再通过 file 命令启动被调试程序,此时被调试程序作为 file 命令的参数传递,命令及回显如下:

```
root@VM-4-7-Ubuntu:/data/code/demo/src#cjdb
(cjdb) file demo
Current executable set to '/data/code/demo/src/demo' (x86_64).
(cjdb)
```

(3) 对于已经处于运行中的程序,可以通过调试器的 attach 命令来附加进程。对于本章的演示代码,因为全局变量 MAX_COUNT 的值比较小,可能一启动就运行结束了,要演示 attach 方式,可以暂时把全局变量 MAX_COUNT 设置成一个较大的值,例如 10 000 000,重新编译后通过 nohup 不挂断的方式启动 demo 程序,根据回显信息记录下 demo 的进程 id,启动 cjdb 后,把进程 id 作为参数传递给命令 attach,这样就把 demo 进程附加到了调成器,命令及回显如下:

```
root@VM-4-7-Ubuntu:/data/code/demo/src#nohup ./demo &
[1] 5657
root@VM-4-7-Ubuntu:/data/code/demo/src# nohup: ignoring input and appending output to
'nohup.out'

root@VM-4-7-Ubuntu:/data/code/demo/src#cjdb
(cjdb) attach 5657
Process 5657 stopped
* thread #1, name = 'demo', stop reason = signal SIGSTOP
    frame #0: 0x00007fee47977a7b
libcangjie-runtime.so`___lldb_unnamed_symbol886 + 155
libcangjie-runtime.so`___lldb_unnamed_symbol886:
-> 0x7fee47977a7b <+155>: movq    %rax, (%rbx)
   0x7fee47977a7e <+158>: movq    %r12, 0x8(%rbx)
   0x7fee47977a82 <+162>: movq    %fs:0x28, %rax
   0x7fee47977a8b <+171>: cmpq    -0x28(%rbp), %rax
  thread #2, name = 'PoolGC_1', stop reason = signal SIGSTOP
    frame #0: 0x00007fee46b9bad3
libpthread.so.0`pthread_cond_wait@@GLIBC_2.3.2 + 579
```

```
libpthread.so.0`pthread_cond_wait@@GLIBC_2.3.2:
  -> 0x7fee46b9bad3 <+579>: cmpq   $-0x1000, %rax       ; imm = 0xF000
     0x7fee46b9bad9 <+585>: movq   0x30(%rsp), %r8
     0x7fee46b9bade <+590>: ja     0x7fee46b9bc10       ; <+896>
     0x7fee46b9bae4 <+596>: movl   %r9d, %edi
   thread #3, name = 'MainGC', stop reason = signal SIGSTOP
     frame #0: 0x00007fee46b9c065
libpthread.so.0`pthread_cond_timedwait@@GLIBC_2.3.2 + 821
libpthread.so.0`pthread_cond_timedwait@@GLIBC_2.3.2:
  -> 0x7fee46b9c065 <+821>: cmpq   $-0x1000, %rax       ; imm = 0xF000
     0x7fee46b9c06b <+827>: ja     0x7fee46b9c20a       ; <+1242>
     0x7fee46b9c071 <+833>: movl   0x50(%rsp), %edi
     0x7fee46b9c075 <+837>: callq  0x7fee46b9f150       ;
__pthread_disable_asynccancel
   thread #4, name = 'FinalProcessor', stop reason = signal SIGSTOP
     frame #0: 0x00007fee46b9c065
libpthread.so.0`pthread_cond_timedwait@@GLIBC_2.3.2 + 821
libpthread.so.0`pthread_cond_timedwait@@GLIBC_2.3.2:
  -> 0x7fee46b9c065 <+821>: cmpq   $-0x1000, %rax       ; imm = 0xF000
     0x7fee46b9c06b <+827>: ja     0x7fee46b9c20a       ; <+1242>
     0x7fee46b9c071 <+833>: movl   0x50(%rsp), %edi
     0x7fee46b9c075 <+837>: callq  0x7fee46b9f150       ;
__pthread_disable_asynccancel
   thread #5, name = 'schmon', stop reason = signal SIGSTOP
     frame #0: 0x00007fee470a9680 libc.so.6`__nanosleep + 64
libc.so.6`__nanosleep:
  -> 0x7fee470a9680 <+64>: cmpq    $-0x1000, %rax       ; imm = 0xF000
     0x7fee470a9686 <+70>: ja      0x7fee470a96b2       ; <+114>
     0x7fee470a9688 <+72>: movl    %edx, %edi
     0x7fee470a968a <+74>: movl    %eax, 0xc(%rsp)
Executable module set to "/data/code/demo/src/demo".
Architecture set to: x86_64-unknown-linux-gnu.
(cjdb)
```

## 27.4 调试命令

### 27.4.1 断点

简单来讲,通常程序是按照顺序一条一条执行指令的,这由程序代码的内在逻辑决定,如果要实现对程序的调试,就需要在被调试的代码行位置中断程序的执行,然后才可以观察或者管理程序的变量和寄存器信息。

仓颉调试器通过设置断点的方式,修改待调试位置的程序指令,把原先要执行的指令替换为中断指令,在程序执行到调试位置后就会引起中断,从而暂停程序的执行。此时调试器把原先要执行的指令恢复,因为程序还在暂停状态,可以进行程序的调试工作,在该位置调试完毕后可以继续向下执行,所以设置断点是调试程序的基础,在仓颉调试器中,有以下5种常用的断点操作指令。

### 1. 设置源码断点

设置源码断点的指令格式如下:

```
breakpoint set --file filename --line line_number
```

其中,filename 是代码文件名称,line_number 是代码行号,假如要对应用 demo 的第 15 行代码添加断点,命令和回显如下:

```
(cjdb) breakpoint set --file Debug_demo.cj --line 15
Breakpoint 1: where = demo`default::main() + 82 at Debug_demo.cj:15:31, address = 0x00000000000383d6
(cjdb)
```

仓颉调试器支持对此指令进行简化,简化后的指令格式如下:

```
b filename : line_number
```

如果只有一个文件,也可以省略文件参数,只需行号,假如要对应用 demo 的第 16 行代码添加断点,简化后的命令和回显如下:

```
(cjdb) b 16
Breakpoint 2: where = demo`default::main() + 95 at Debug_demo.cj:16:9, address = 0x00000000000383e3
```

### 2. 设置函数断点

设置函数断点的指令格式如下:

```
breakpoint set --name function_name
```

也可以简写为

```
b function_name
```

假如要对应用 demo 的 addRandomNum 函数添加断点,简化后的命令和回显如下:

```
(cjdb) b addRandomNum
Breakpoint 3: where = demo`default::addRandomNum(Int64) + 53 at Debug_demo.cj:24:16, address = 0x0000000000005d19
```

### 3. 设置条件断点

条件断点不是每次执行都触发,只有在满足条件时才会触发中断,指令格式如下:

```
breakpoint set --file filename --line line_number --condition expression
```

其中,filename 是代码文件名称,line_number 是代码行号,expression 为触发条件。目前触发条件只支持基础类型的变量条件设置(包括 Int8、Int16、Int32、Int64、UInt8、UInt16、UInt32、UInt64、Bool、Char),条件运算符支持==、!=、>、<、>=、<=、and、or 运算符。该指令的简写格式如下:

```
b -f filename -l line_number -c expression
```

假如要对应用 demo 的第 15 行代码添加断点,并且在变量 sumNum 等于 10 时触发,简化的命令和回显如下:

```
(cjdb) b -l 15 -c 'sumNum == 10'
Breakpoint 4: where = demo`default::main() + 81 at Debug_demo.cj:15:31, address = 0x0000000000601a95
```

因为应用只有一个代码文件,上述命令省略了-f 参数。

#### 4. 显示所有断点

显示当前所有断点的指令格式如下:

```
breakpoint list
```

也可以简写为

```
br l
```

使用该命令后,将列出当前所有的断点信息:

```
(cjdb) br l
Current breakpoints:
1: file = 'Debug_demo.cj', line = 15, exact_match = 0, locations = 1
  1.1: where = demo`default::main() + 82 at Debug_demo.cj:15:31, address = demo[0x00000000000383d6], unresolved, hit count = 0

2: file = '/data/code/demo/src/Debug_demo.cj', line = 16, exact_match = 0, locations = 1
  2.1: where = demo`default::main() + 95 at Debug_demo.cj:16:9, address = demo[0x00000000000383e3], unresolved, hit count = 0

3: name = 'addRandomNum', locations = 1
  3.1: where = demo`default::addRandomNum(Int64) + 53 at Debug_demo.cj:24:16, address = demo[0x0000000000005d19], unresolved, hit count = 0

4: file = '/data/code/demo/src/Debug_demo.cj', line = 15, exact_match = 0, locations = 1
Condition: sumNum == 10

  4.1: where = demo`default::main() + 82 at Debug_demo.cj:15:31, address = demo[0x00000000000383d6], unresolved, hit count = 0
```

#### 5. 删除断点

仓颉调试器可以删除断点,指令格式如下:

```
breakpoint delete
```

执行指令后,可以根据提示一次性删除所有断点。

删除断点的指令可以简写为

```
br del
```

如果要删除特定的断点,则可以在指令后面添加要删除的断点编号,多个断点编号可以用空格分隔,示例命令和回显如下:

```
(cjdb) br del 2 3
2 breakpoints deleted; 0 breakpoint locations disabled.
```

## 27.4.2 观察点

当内存中一个指定的地址被读、写或者修改时就暂停的断点被称为观察点,观察点是一种特殊的断点,又称为数据断点。

**1. 设置观察点**

设置观察点的指令格式如下:

```
watchpoint set variable -w point_type variable_name
```

其中,-w 表示观察点类型,point_type 为 read、write、read_write 3 种类型之一；variable_name 表示变量的名称,目前只支持在基础类型设置观察点。

该指令的缩写形式如下:

```
wa s v -w point_type variable_name
```

假如要对应用 demo 的 sumNum 变量添加读观察点,简化的命令和回显如下:

```
(cjdb) wa s v -w read sumNum
Watchpoint created: Watchpoint 1: addr = 0x7fffd5632eb0 size = 8 state = enabled type = r
    declare @ '/data/code/demo/src/Debug_demo.cj:11'
    watchpoint spec = 'sumNum'
    new value: 0
```

需要注意的是,添加观察点时需要被调试程序已经处于启动状态,否则添加观察点时会出现 error: Command requires a current process 的错误信息。

**2. 显示所有观察点**

显示当前所有观察点的指令格式如下:

```
watchpoint list
```

也可以简写为

```
wa l
```

使用该命令后,将列出当前所有的观察点信息:

```
(cjdb) wa l
Number of supported hardware watchpoints: 4
Current watchpoints:
Watchpoint 1: addr = 0x7fffd5632eb0 size = 8 state = enabled type = r
    declare @ '/data/code/demo/src/Debug_demo.cj:11'
    watchpoint spec = 'sumNum'
    new value: 0
```

**3. 删除观察点**

删除观察点的指令格式如下:

```
watchpoint delete
```

执行指令后,可以根据提示一次性删除所有观察点。

删除观察点的指令可以简写为

```
wa del
```

如果要删除特定的观察点,则可以在指令后面添加要删除的观察点编号,多个观察点编号可以用空格分隔,示例命令和回显如下:

```
(cjdb) wa del 1 2
2 watchpoints deleted.
```

### 27.4.3 启动

启动被调试的程序指令为 r,执行该指令后,程序会停留在第 1 个遇到的断点位置,假如已经在第 11 行添加了断点,执行 r 指令的回显如下:

```
(cjdb) r
Process 13811 launched: '/data/code/demo/src/demo' (x86_64)
Process 13811 stopped
* thread #1, name = 'demo', stop reason = breakpoint 1.1
    frame #0: 0x000055555558c3c1 demo`default::main() at Debug_demo.cj:11:5
      8         var index = 0
      9
      10        //累加值
   -> 11        var sumNum = 0
      12
      13        //计算 MAX_COUNT 个随机值的累加值
      14        while (index < MAX_COUNT) {
```

### 27.4.4 执行

调试器的主要执行指令包括 n(next)、c(continue)、s(stepin)、finish 几个,下面分别通过示例说明。

#### 1. 单步执行

单步执行的指令是 next,简写为 n,将程序执行到下一行代码,示例如下:

```
(cjdb) n
Process 20956 stopped
* thread #1, name = 'demo', stop reason = step over
    frame #0: 0x000055555558c3d6 demo`default::main() at Debug_demo.cj:15:31
      12
      13        //计算 MAX_COUNT 个随机值的累加值
      14        while (index < MAX_COUNT) {
   -> 15            sumNum = addRandomNum(sumNum)
      16            index++
      17        }
      18        //打印累加值
(cjdb) n
```

```
Process 20956 stopped
* thread #1, name = 'demo', stop reason = step over
    frame #0: 0x000055555558c3e3 demo`default::main() at Debug_demo.cj:16:9
   13           //计算 MAX_COUNT 个随机值的累加值
   14           while (index < MAX_COUNT) {
   15               sumNum = addRandomNum(sumNum)
-> 16               index++
   17           }
   18           //打印累加值
   19           println(sumNum)
```

第 1 个指令 n 执行到了第 15 行代码，第 2 个指令 n 就执行到了第 16 行。

### 2．执行到下一断点

执行到下一断点的指令是 continue，简写为 c，执行程序到下一个断点位置，示例如下：

```
Process 21670 stopped
* thread #1, name = 'demo', stop reason = breakpoint 1.1 2.1
    frame #0: 0x000055555558c3c1 demo`default::main() at Debug_demo.cj:11:5
   8        var index = 0
   9
   10       //累加值
-> 11       var sumNum = 0
   12
   13       //计算 MAX_COUNT 个随机值的累加值
   14       while (index < MAX_COUNT) {
(cjdb) c
Process 21670 resuming
Process 21670 stopped
* thread #1, name = 'demo', stop reason = breakpoint 3.1
    frame #0: 0x000055555558c3d6 demo`default::main() at Debug_demo.cj:15:31
   12
   13           //计算 MAX_COUNT 个随机值的累加值
   14           while (index < MAX_COUNT) {
-> 15               sumNum = addRandomNum(sumNum)
   16               index++
   17           }
   18           //打印累加值
(cjdb)
```

被调试程序停在了第 11 行位置，然后输入指令 c，程序就执行到了下一个断点位置，也就是第 15 行。

### 3．函数进入

函数进入的指令是 s，在程序遇到函数调用时，该指令可以进入被调用函数的声明处，示例如下：

```
Process 21670 stopped
* thread #1, name = 'demo', stop reason = breakpoint 3.1
    frame #0: 0x000055555558c3d6 demo`default::main() at Debug_demo.cj:15:31
```

```
         12
         13          //计算 MAX_COUNT 个随机值的累加值
         14          while (index < MAX_COUNT) {
  -> 15                  sumNum = addRandomNum(sumNum)
         16              index++
         17          }
         18          //打印累加值
(cjdb) s
Process 21670 stopped
* thread #1, name = 'demo', stop reason = step in
    frame #0: 0x0000555555559d19 demo`default::addRandomNum(currentSum = 0) at Debug_demo.cj:24:16
         21
         22          //产生一个随机数和参数 currentSum 相加,然后返回相加的和
         23          func addRandomNum(currentSum: Int64): Int64 {
  -> 24              let rand = Random()
         25              let newSum = currentSum + rand.nextInt64(100)
         26              return newSum
         27          }
```

被调试程序停留在了第 15 行,该行调用了函数 addRandomNum,通过指令 s 就进入了该函数内部,从而可以在函数 addRandomNum 内继续进行调试。

### 4. 函数退出

函数退出的指令是 finish,该指令用于退出当前函数的调试,返回上一个调用栈,示例如下:

```
Process 21670 stopped
* thread #1, name = 'demo', stop reason = breakpoint 3.1
    frame #0: 0x000055555558c3d6 demo`default::main() at Debug_demo.cj:15:31
         12
         13          //计算 MAX_COUNT 个随机值的累加值
         14          while (index < MAX_COUNT) {
  -> 15                  sumNum = addRandomNum(sumNum)
         16              index++
         17          }
         18          //打印累加值
(cjdb) s
Process 21670 stopped
* thread #1, name = 'demo', stop reason = step in
    frame #0: 0x0000555555559d19 demo`default::addRandomNum(currentSum = 131)
at Debug_demo.cj:24:16
         21
         22          //产生一个随机数和参数 currentSum 相加,然后返回相加的和
         23          func addRandomNum(currentSum: Int64): Int64 {
  -> 24              let rand = Random()
         25              let newSum = currentSum + rand.nextInt64(100)
         26              return newSum
         27          }
```

```
(cjdb) finish
Process 21670 stopped
* thread #1, name = 'demo', stop reason = step out
Return value: (Int64) $0 = 156

    frame #0: 0x000055555558c3df demo`default::main() at Debug_demo.cj:15:9
   12
   13           //计算 MAX_COUNT 个随机值的累加值
   14           while (index < MAX_COUNT) {
-> 15               sumNum = addRandomNum(sumNum)
   16               index++
   17           }
   18           //打印累加值
(cjdb)
```

在这个示例里,先使用指令 s 进入了被调用的函数内部,然后使用指令 finish 退出函数,回到上一个调用栈。

### 27.4.5 变量

程序调试中,查看变量的值是判断程序执行是否正确的重要手段,仓颉调试器提供了查看和修改变量值的指令,下面分别进行说明。

**1. 查看单个变量值**

查看单个变量值的指令格式如下:

```
print variable_name
```

其中,variable_name 为变量名称,该指令可以简写为

```
p variable_name
```

示例如下:

```
Process 21670 stopped
* thread #1, name = 'demo', stop reason = step over
    frame #0: 0x000055555558c3e3 demo`default::main() at Debug_demo.cj:16:9
   13           //计算 MAX_COUNT 个随机值的累加值
   14           while (index < MAX_COUNT) {
   15               sumNum = addRandomNum(sumNum)
-> 16               index++
   17           }
   18           //打印累加值
   19           println(sumNum)
(cjdb) print index
(Int64) $3 = 3
(cjdb) p sumNum
(Int64) $4 = 202
(cjdb) p MAX_COUNT
(Int64) $5 = 100
(cjdb)
```

在上例中,分别查看了局部变量 index、sumNum 和全局变量 MAX_COUNT 的值(MAX_COUNT 的值和源代码有关,如果源代码里是 100,则这里也是 100)。

### 2. 批量查看局部变量值

除了查看单个变量值,仓颉调试器支持批量查看,使用 locals(或 local)可以查看断点所在函数生命周期内的所有局部变量值,示例如下:

```
Process 21670 stopped
* thread #1, name = 'demo', stop reason = breakpoint 3.1
    frame #0: 0x000055555558c3d6 demo`default::main() at Debug_demo.cj:15:31
   12
   13          //计算 MAX_COUNT 个随机值的累加值
   14          while (index < MAX_COUNT) {
-> 15              sumNum = addRandomNum(sumNum)
   16              index++
   17          }
   18          //打印累加值
(cjdb) locals
(Int64) index = 4
(Int64) sumNum = 202
```

### 3. 批量查看全局变量值

查看所有全局变量值的指令为 globals(或 global),示例如下:

```
Process 21670 stopped
* thread #1, name = 'demo', stop reason = breakpoint 3.1
    frame #0: 0x000055555558c3d6 demo`default::main() at Debug_demo.cj:15:31
   12
   13          //计算 MAX_COUNT 个随机值的累加值
   14          while (index < MAX_COUNT) {
-> 15              sumNum = addRandomNum(sumNum)
   16              index++
   17          }
   18          //打印累加值
(cjdb) globals
Global variables for /data/code/demo/src/default in
/data/code/demo/src/demo:
(Int64) default::MAX_COUNT = 100
(cjdb)
```

### 4. 修改变量值

修改变量值的指令为 set,可以修改局部变量和全局变量的值,示例如下:

```
Process 21670 stopped
* thread #1, name = 'demo', stop reason = breakpoint 3.1
    frame #0: 0x000055555558c3d6 demo`default::main() at Debug_demo.cj:15:31
   12
   13          //计算 MAX_COUNT 个随机值的累加值
   14          while (index < MAX_COUNT) {
-> 15              sumNum = addRandomNum(sumNum)
   16              index++
```

```
           17          }
           18          //打印累加值
(cjdb) local
(Int64) index = 6
(Int64) sumNum = 320
(cjdb) global
Global variables for /data/code/demo/src/default in
/data/code/demo/src/demo:
(Int64) default::MAX_COUNT = 100
(cjdb) set index = 10
(Int64) $6 = 10
(cjdb) set MAX_COUNT = 200
(Int64) $7 = 200
(cjdb) c
Process 21670 resuming
Process 21670 stopped
* thread #1, name = 'demo', stop reason = breakpoint 3.1
    frame #0: 0x000055555558c3d6 demo`default::main() at Debug_demo.cj:15:31
           12
           13          //计算 MAX_COUNT 个随机值的累加值
           14          while (index < MAX_COUNT) {
->  15              sumNum = addRandomNum(sumNum)
           16              index++
           17          }
           18          //打印累加值
(cjdb) local
(Int64) index = 11
(Int64) sumNum = 383
(cjdb) global
Global variables for /data/code/demo/src/default in
/data/code/demo/src/demo:
(Int64) default::MAX_COUNT = 200
(cjdb)
```

在上述示例中，先修改了变量 index 和 MAX_COUNT 的值，然后运行到下一个断点，继续查看变量值，可以看到程序已经按照修改后的变量值运行了。

**5. 自定义类型示例**

除了基本数据类型，仓颉调试器支持对自定义类型的变量进行查看和修改，例如 Class、Record、Array 等，下面通过一个示例演示如何对自定义类型的变量值进行查看和修改，示例代码如下：

```
//Chapter27/set_var_demo.cj

class Rectangle {
    Rectangle(var width: Int64, var height: Int64) {}
    func area() {
        return width * height
    }
}
```

```
main() {
    let rectArray = Array<Rectangle>([Rectangle(2, 10), Rectangle(100, 10)])

    for (item in rectArray) {
        println(item.area())
    }
}
```

对上述示例程序进行调试,断点可以设置在第 12 行 println(item.area()) 位置,调试示例如下:

```
Process 26560 stopped
 * thread #1, name = 'demo', stop reason = breakpoint 1.1
     frame #0: 0x0000555555580d81 demo`default::main() at set_var_demo.cj:12:17
   9          let rectArray = Array<Rectangle>([Rectangle(2, 10), Rectangle(100, 10)])
   10
   11         for (item in rectArray) {
-> 12             println(item.area())
   13         }
   14     }
   15
(cjdb) print rectArray
(Array<Rectangle>) $0 = {
  [0] = {
    width = 2
    height = 10
  }
  [1] = {
    width = 100
    height = 10
  }
}
(cjdb) print item
(Rectangle) $1 = {
  width = 2
  height = 10
}
(cjdb) set rectArray[1].width = 800
(Int64) $2 = 800
(cjdb)
```

在上述示例中,打印了变量 rectArray 和 item 的值,并将 rectArray 第 2 个元素的成员 width 值设置为 800。下面,输入指令 c 继续执行,然后打印变量 item 的值:

```
(cjdb) c
Process 26560 resuming
20
Process 26560 stopped
 * thread #1, name = 'demo', stop reason = breakpoint 1.1
     frame #0: 0x0000555555580d81 demo`default::main() at set_var_demo.cj:12:17
```

```
     9              let rectArray = Array<Rectangle>([Rectangle(2, 10), Rectangle(100, 10)])
    10
    11              for (item in rectArray) {
 -> 12                  println(item.area())
    13              }
    14          }
    15
(cjdb) p item
(Rectangle) $3 = {
  width = 800
  height = 10
}
(cjdb)
```

可以看到，上一步对 Rectangle 成员变量 width 值的修改已经生效。

## 27.4.6　退出

退出调试的指令为 exit，如果被调试程序还没有运行，则该指令会让调试器直接退出，否则会给出提示：

```
(cjdb) exit
Quitting LLDB will kill one or more processes. Do you really want to proceed: [Y/n]
```

输入 y 会退出调试器。

# 图书推荐

| 书 名 | 作 者 |
|---|---|
| 仓颉语言核心编程——入门、进阶与实战 | 徐礼文 |
| 仓颉语言程序设计 | 董昱 |
| 仓颉程序设计语言 | 刘安战 |
| 仓颉语言极速入门——UI全场景实战 | 张云波 |
| HarmonyOS 移动应用开发（ArkTS版） | 刘安战、余雨萍、陈争艳 等 |
| 深度探索 Vue.js——原理剖析与实战应用 | 张云鹏 |
| 前端三剑客——HTML5＋CSS3＋JavaScript 从入门到实战 | 贾志杰 |
| 剑指大前端全栈工程师 | 贾志杰、史广、赵东彦 |
| Flink 原理深入与编程实战——Scala＋Java（微课视频版） | 辛立伟 |
| Spark 原理深入与编程实战（微课视频版） | 辛立伟、张帆、张会娟 |
| PySpark 原理深入与编程实战（微课视频版） | 辛立伟、辛雨桐 |
| HarmonyOS 应用开发实战（JavaScript 版） | 徐礼文 |
| HarmonyOS 原子化服务卡片原理与实战 | 李洋 |
| 鸿蒙操作系统开发入门经典 | 徐礼文 |
| 鸿蒙应用程序开发 | 董昱 |
| 鸿蒙操作系统应用开发实践 | 陈美汝、郑森文、武延军、吴敬征 |
| HarmonyOS 移动应用开发 | 刘安战、余雨萍、李勇军 等 |
| HarmonyOS App 开发从 0 到 1 | 张诏添、李凯杰 |
| JavaScript 修炼之路 | 张云鹏、戚爱斌 |
| JavaScript 基础语法详解 | 张旭乾 |
| 华为方舟编译器之美——基于开源代码的架构分析与实现 | 史宁宁 |
| Android Runtime 源码解析 | 史宁宁 |
| 恶意代码逆向分析基础详解 | 刘晓阳 |
| 网络攻防中的匿名链路设计与实现 | 杨昌家 |
| 深度探索 Go 语言——对象模型与 runtime 的原理、特性及应用 | 封幼林 |
| 深入理解 Go 语言 | 刘丹冰 |
| Vue＋Spring Boot 前后端分离开发实战 | 贾志杰 |
| Spring Boot 3.0 开发实战 | 李西明、陈立为 |
| Vue.js 光速入门到企业开发实战 | 庄庆乐、任小龙、陈世云 |
| Flutter 组件精讲与实战 | 赵龙 |
| Flutter 组件详解与实战 | ［加］王浩然（Bradley Wang） |
| Dart 语言实战——基于 Flutter 框架的程序开发（第 2 版） | 亢少军 |
| Dart 语言实战——基于 Angular 框架的 Web 开发 | 刘仕文 |
| IntelliJ IDEA 软件开发与应用 | 乔国辉 |
| Python 量化交易实战——使用 vn.py 构建交易系统 | 欧阳鹏程 |
| Python 从入门到全栈开发 | 钱超 |
| Python 全栈开发——基础入门 | 夏正东 |
| Python 全栈开发——高阶编程 | 夏正东 |
| Python 全栈开发——数据分析 | 夏正东 |

续表

| 书　名 | 作　者 |
|---|---|
| Python 编程与科学计算（微课视频版） | 李志远、黄化人、姚明菊 等 |
| Diffusion AI 绘图模型构造与训练实战 | 李福林 |
| 图像识别——深度学习模型理论与实战 | 于浩文 |
| 数字 IC 设计入门（微课视频版） | 白栎旸 |
| 动手学推荐系统——基于 PyTorch 的算法实现（微课视频版） | 於方仁 |
| 人工智能算法——原理、技巧及应用 | 韩龙、张娜、汝洪芳 |
| Python 数据分析实战——从 Excel 轻松入门 Pandas | 曾贤志 |
| Python 概率统计 | 李爽 |
| Python 数据分析从 0 到 1 | 邓立文、俞心宇、牛瑶 |
| 从数据科学看懂数字化转型——数据如何改变世界 | 刘通 |
| 鲲鹏架构入门与实战 | 张磊 |
| 鲲鹏开发套件应用快速入门 | 张磊 |
| 华为 HCIA 路由与交换技术实战 | 江礼教 |
| 华为 HCIP 路由与交换技术实战 | 江礼教 |
| openEuler 操作系统管理入门 | 陈争艳、刘安战、贾玉祥 等 |
| 5G 核心网原理与实践 | 易飞、何宇、刘子琦 |
| Python 游戏编程项目开发实战 | 李志远 |
| 编程改变生活——用 Python 提升你的能力（基础篇·微课视频版） | 邢世通 |
| 编程改变生活——用 Python 提升你的能力（进阶篇·微课视频版） | 邢世通 |
| 编程改变生活——用 PySide6/PyQt6 创建 GUI 程序（基础篇·微课视频版） | 邢世通 |
| 编程改变生活——用 PySide6/PyQt6 创建 GUI 程序（进阶篇·微课视频版） | 邢世通 |
| FFmpeg 入门详解——音视频原理及应用 | 梅会东 |
| FFmpeg 入门详解——SDK 二次开发与直播美颜原理及应用 | 梅会东 |
| FFmpeg 入门详解——流媒体直播原理及应用 | 梅会东 |
| FFmpeg 入门详解——命令行与音视频特效原理及应用 | 梅会东 |
| FFmpeg 入门详解——音视频流媒体播放器原理及应用 | 梅会东 |
| 精讲 MySQL 复杂查询 | 张方兴 |
| Python Web 数据分析可视化——基于 Django 框架的开发实战 | 韩伟、赵盼 |
| Python 玩转数学问题——轻松学习 NumPy、SciPy 和 Matplotlib | 张骞 |
| Pandas 通关实战 | 黄福星 |
| 深入浅出 Power Query M 语言 | 黄福星 |
| 深入浅出 DAX——Excel Power Pivot 和 Power BI 高效数据分析 | 黄福星 |
| 从 Excel 到 Python 数据分析：Pandas、xlwings、openpyxl、Matplotlib 的交互与应用 | 黄福星 |
| 云原生开发实践 | 高尚衡 |
| 云计算管理配置与实战 | 杨昌家 |
| 虚拟化 KVM 极速入门 | 陈涛 |
| 虚拟化 KVM 进阶实践 | 陈涛 |
| HarmonyOS 从入门到精通 40 例 | 戈帅 |
| OpenHarmony 轻量系统从入门到精通 50 例 | 戈帅 |
| AR Foundation 增强现实开发实战（ARKit 版） | 汪祥春 |
| AR Foundation 增强现实开发实战（ARCore 版） | 汪祥春 |